Springer Undergraduate Mathematics Series

Springer
London
Berlin
Heidelberg
New York
Barcelona
Hong Kong
Milan
Paris
Santa Clara
Singapore
Tokyo

Other books in this series

Analytic Methods for Partial Differential Equations *G. Evans, J. Blackledge, P. Yardley*
Applied Geometry for Computer Graphics and CAD *D. Marsh*
Basic Linear Algebra *T.S. Blyth and E.F. Robertson*
Basic Stochastic Processes *Z. Brzeźniak and T. Zastawniak*
Elements of Logic via Numbers and Sets *D.L. Johnson*
Elementary Number Theory *G.A. Jones and J.M. Jones*
Groups, Rings and Fields *D.A.R. Wallace*
Hyperbolic Geometry *J.W. Anderson*
Introduction to Laplace Transforms and Fourier Series *P.P.G. Dyke*
Introduction to Ring Theory *P.M. Cohn*
Introductory Mathematics: Algebra and Analysis *G. Smith*
Introductory Mathematics: Applications and Methods *G.S. Marshall*
Measure, Integral and Probability *M. Capiński and E. Kopp*
Multivariate Calculus and Geometry *S. Dineen*
Numerical Methods for Partial Differential Equations *G. Evans, J. Blackledge, P. Yardley*
Sets, Logic and Categories *P.J. Cameron*
Topologies and Uniformities *I.M. James*
Vector Calculus *P.C. Matthews*

P.M. Cohn

Introduction to Ring Theory

Springer

Paul M. Cohn, MA, PhD, FRS
Department of Mathematics, University College London, Gower Street,
London WC1E 6BT, UK

Cover illustration elements reproduced by kind permission of:
Aptech Systems, Inc., Publishers of the GAUSS Mathematical and Statistical System, 23804 S.E. Kent-Kangley Road, Maple Valley, WA 98038, USA. Tel: (206) 432 - 7855 Fax (206) 432 - 7832 email: info@aptech.com.URL: www.aptech.com
American Statistical Association: Chance Vol 8 No 1, 1995 article by KS and KW Heiner 'Tree Rings of the Northern Shawangunks' page 32 fig 2
Springer-Verlag: Mathematica in Education and Research Vol 4 Issue 3 1995 article by Roman E Maeder, Beatrice Amrhein and Oliver Gloor 'Illustrated Mathematics: Visualization of Mathematical Objects' page 9 fig 11, originally published as a CD ROM 'Illustrated Mathematics' by TELOS: ISBN 0-387-14222-3, German edition by Birkhauser: ISBN 3-7643-5100-4.
Mathematics in Education and Research Vol 4 Issue 3 1995 article by Richard J Gaylord and Kazume Nishidate 'Traffic Engineering with Cellular Automata' page 35 fig 2. Mathematica in Education and Research Vol 5 Issue 2 1996 article by Michael Trott 'The Implicitization of a Trefoil Knot' page 14.
Mathematica in Education and Research Vol 5 Issue 2 1996 article by Lee de Cola 'Coins, Trees, Bars and Bells: Simulation of the Binomial Process page 19 fig 3. Mathematica in Education and Research Vol 5 Issue 2 1996 article by Richard Gaylord and Kazume Nashidate 'Contagious Spreading' page 33 fig 1. Mathematica in Education and Research Vol 5 Issue 2 1996 article by Joe Buhler and Stan Wagon 'Secrets of the Madelung Constant' page 50 fig 1.

ISBN 1-85233-206-9 Springer-Verlag London Berlin Heidelberg

British Library Cataloguing in Publication Data
Cohn, P. M. (Paul Moritz)
Introduction to ring theory. – (Springer undergraduate mathematics series)
1. Rings (Algebra)
I. Title
512.4
ISBN 1852332069

Library of Congress Cataloging-in-Publication Data
Cohn, P. M. (Paul Moritz)
Introduction to ring theory/P. M. Cohn.
p. cm. – (Springer undergraduate mathematics series)
Includes bibliographical references and index.
ISBN 1–85233–206–9 (alk. paper)
1. Rings (Algebra) I. Title. II. Series.
QA247.C637 2000 99-41609
512′.4–dc21 CIP

Printed in Great Britain
2nd printing 2001

Typesetting by BC Typesetting, Bristol BS31 1NZ
Printed and bound at the Athenæum Press Ltd., Gateshead, Tyne & Wear
12/3830-54321 Printed on acid-free paper SPIN 10834972

For my grandchildren
James, Olivia and Hugo Aaronson

Preface

The theory of rings, a newcomer on the algebra scene in the late 19th century, has grown in importance throughout the 20th century, and is now a well-established part of algebra. This introduction to the subject is intended for second and third year students and postgraduates, but would also be suitable for self-study. The reader is assumed to have met sets, groups and vector spaces, but any results used are clearly stated, with a reference to sources.

After a chapter on the basic definitions of rings, modules and a few categorical notions, there is a chapter each on Artinian rings and on (commutative) Noetherian rings, followed by a presentation of some important constructions and concepts in ring theory, such as direct products, tensor products and the use of projective and injective modules. Of the many possible applications we have chosen to include at least a brief account of representation theory and algebraic number theory.

In the final chapter an author is usually allowed to indulge his personal tastes; here it is devoted to free algebras and more generally, free ideal rings. These are topics that have only recently been developed, whose basic theory is relatively simple, and which are not too widely known, although they are now beginning to find applications in topology and functional analysis.

Care has been taken to present all proofs in detail, while including motivation and providing material for the reader to practise on. Most results are accompanied by worked examples and each section has a number of exercises, for whose solution some hints are given at the end of the book.

I would like to express my thanks to the Publishers, to their Mathematics Editor Mr. D. Ireland and especially to their Copy Editor Mr S.M. Nugent for the expert way in which he has helped to tame the manuscript. I would also like to thank an anonymous reviewer for bringing a number of mistakes in the original manuscript to my attention.

September 1999 P. M. Cohn

Contents

Introduction

Most parts of algebra have undergone great changes in the course of the 20th century; probably none more so than the theory of rings. While groups were well established early in the 19th century, in its closing years the term "ring" just meant a "ring of algebraic integers" and the axiomatic foundations were not laid until 1914 (Fraenkel 1914). In the early years of the 20th century Wedderburn proved his theorems about finite-dimensional algebras, which some time later were recognized by Emmy Noether to be true of rings with chain condition; Emil Artin realized that the minimum condition was enough and such rings are now called Artinian. Both E. Noether and E. Artin lectured on algebra in Göttingen in the 1920s and B.L. van der Waerden, who had attended these lectures, decided to write a textbook based on them which first came out in 1931 and rapidly became a classic. Significantly enough it was entitled "Moderne Algebra" on its first appearance, but the "Modern" was dropped from the title after 25 years.

Meanwhile in commutative ring theory the properties of rings of algebraic numbers and coordinate rings of algebraic varieties were established by E. Noether and others for more general rings: apart from commutativity only the maximum condition was needed and the rings became known as Noetherian rings.

Whereas the trend in mathematics at the beginning of the 20th century was towards axiomatics, the emphasis since the second world war has been on generality, where possible, regarding several subjects from a single point of view. Thus the algebraicization of topology has led to the development of homological algebra, which in turn has had a major influence on ring theory.

In another field, the remarkable progress in algebraic geometry by the Italian School has been put on a firm algebraic basis, and this has led to progress in commutative ring theory, culminating in the result which associates with any commutative ring an affine scheme. Thirdly the theory of operator algebras, which itself received an impetus from quantum mechanics, has led to the development of function algebras. More recently the study of "non-commutative geometry" has emphasized the role of non-commutative rings.

To do justice to all these trends would require several volumes and many authors. The present volume merely takes the first steps in such a development, introducing the definition and basic notions of rings and constructions, such as rings of fractions, residue-class rings and matrix rings, to be followed by brief accounts of Artinian rings, commutative Noetherian rings, and a chapter on free rings.

Suggestions for further reading are included in the Bibliography at the end of the book. References to the author's three-volume work on Algebra are abbreviated to A.1, A.2, and A.3, when cited in the text.

Remarks on Notation and Terminology

Most of the notation will be explained when it first occurs, but we assume that such ideas as the set of natural numbers $\mathbb{N} = \{0, 1, 2, \ldots\}$, integers $\mathbb{Z}$, rational numbers $\mathbb{Q}$, and real numbers $\mathbb{R}$, are known, as well as the membership sign: $\in$, as in $3 \in \mathbb{N}$, inclusion $\subseteq$ as in $\mathbb{N} \subseteq \mathbb{Z}$, or even $\mathbb{N} \subset \mathbb{Z}$, where "$\subset$" denotes proper inclusion; one also writes $\mathbb{Z} \supseteq \mathbb{N}$, $\mathbb{Z} \supset \mathbb{N}$. This notation for the various systems was introduced by Bourbaki in the late 1930s and is now used worldwide; until that time, surprisingly, no universally used names existed. Instead of $\mathbb{N}$ one sometimes writes $\mathbb{Z}^+$ (the integers ≥ 0). Similarly, $\mathbb{Q}^+$, $\mathbb{R}^+$ indicates the non-negative rational, resp., real numbers. Since every non-zero real number has a positive square, equations such as $x^2 + 1 = 0$ have no real solution, but by introducing a solution i, one obtains the set of all complex numbers $x + y\mathrm{i}(x, y$ real), usually denoted by $\mathbb{C}$. Then one can show that every polynomial equation with complex coefficients has a complex root, a fact expressed by saying that $\mathbb{C}$ is **algebraically closed.**

Of course we assume that the reader is familiar with the notion of a **set**, as a collection of objects, called **elements** or **members** of the set. Given sets S, T, their intersection is written $S \cap T$, their union is $S \cup T$, and their direct product, consisting of all pairs (x, y), where $x \in S, y \in T$, is $S \times T$. Sometimes a slight variant is needed: suppose we have a relation between vectors x_1, x_2, x_3 in a vector space, e.g. $a_1x_1 + a_2x_2 + a_3x_3 = 0$; we may wish to consider the numbers a_1, a_2, a_3, but if $a_1 = a_2$ say, the set $\{a_1, a_2, a_3\}$ will reduce to $\{a_1, a_3\}$ and so will not distinguish between a_1 and a_2. Instead we shall speak of the **family** (a_1, a_2, a_3), **indexed** by the set $\{1, 2, 3\}$.

As a rule we shall use Greek letters for general indexing sets, and Latin letters for finite sets. Sometimes a set with no elements is considered; this is the **empty set**, denoted by $\varnothing$; this notation was introduced by Lefschetz in 1942 (Solomon Lefschetz, 1884–1972). A property which holds for all except a finite number of members of a set S is said to hold for **almost all** members of S. If S' is a subset of a set S, its **complement,** i.e. the set $\{x \in S | x \notin S'\}$, is denoted by $S \setminus S'$.

Our readers will also have met groups before, but for completeness we include their definition here. By a **group** we understand a non-empty set G with a binary operation, called **multiplication,** associating with each pair of elements a, b of G another element of G, written $a.b$ or ab and called the **product** of a and b, such that:

G.1 $a(bc) = (ab)c$; (associative law)

G.2 for any $a, b \in G$ the equations $ax = b, ya = b$ each have a solution.

It turns out that the solutions x, y in G.2 are uniquely determined by a and b. The element e satisfying $ae = a$ also satisfies $ea = a$ for all $a \in G$ and is called the **unit element** or **neutral element** of G. If only G.1 holds, we have a set with an associative multiplication; this is called a **semigroup.** A semigroup with a neutral element is called a **monoid.**

Most of our sets will be finite; those that are infinite will usually be **countable,** i.e. the elements can be enumerated by means of the natural numbers $\mathbb{N}$. No doubt all readers are aware that there are uncountable sets (e.g. the real numbers $\mathbb{R}$), but their properties will not concern us here, though in Chapter 4 we shall briefly deal with this more general situation.

We shall frequently use the notion of a partial ordering on a set and so briefly recall the definition. A **partially ordered** set is a set S with a binary relation $\leq$ satisfying the following conditions:

O.1 for all $x \in S, x \leq x$ (reflexivity);

O.2 for all $x, y, z \in S, x \leq y, y \leq z$ implies $x \leq z$ (transitivity);

O.3 for all $x, y \in S, x \leq y, y \leq x$ implies $x = y$ (antisymmetry).

We shall also write "$x < y$" for "$x \leq y$ and $x \neq y$" and instead of "$x \leq y$" sometimes write "$y \geq x$"; similarly "$x > y$" stands for "$y < x$". If in addition, we have

O.4 for all $x, y \in S, x \leq y$ or $y \leq x$,

the ordering is said to be **total** or **linear**. In any partially ordered set a totally ordered subset is called a **chain**.

Ordered sets abound in mathematics; for example, the natural numbers form a set which is totally ordered by size and partially (but not totally) ordered by divisibility: $a|b$ if and only if $b = ac$ for some integer c. Another example is the collection $\mathscr{P}(S)$ of all subsets of a given set S; it is partially ordered by inclusion, but not totally ordered, unless S has at most one element. If a, b are members of a totally ordered set, then $\max\{a, b\}$ denotes the larger and $\min\{a, b\}$ the smaller of a and b; similarly for finite sets of more than two elements.

In any partially ordered set S, a **greatest** element is an element c such that $x \leq c$ for all $x \in S$; if c is such that $c < x$ for no x, then c is said to be **maximal**. Clearly a greatest element is maximal, but not necessarily conversely. Least and minimal elements are defined similarly. If T is a subset of S, then an **upper bound** of T in S is an element $c \in S$ such that $t \leq c$ for all $t \in T$; lower bounds are defined dually.

Possibly even more basic is the notion of equivalence. By an **equivalence relation** on a set S is meant a binary relation $\sim$ on S such that the following hold:

E.1 for all $x \in S, x \sim x$ (reflexivity);

E.2 for all $x, y \in S, x \sim y \Rightarrow y \sim x$ (symmetry);

E.3 for all $x, y, z \in S, x \sim y$ and $y \sim z \Rightarrow x \sim z$ (transitivity).

It is easily seen that an equivalence relation on S divides S up into blocks, where each element of S lies in exactly one block and two elements belong to the same block if and only if they are equivalent.

Another notion that is basic for all of algebra is that of a mapping. We are familiar with the notion of a function such as x^2, which associates with any real number x its square x^2. This represents the simplest case of a mapping, from $\mathbb{R}$ to $\mathbb{R}$. Generally, a mapping from a set S to a set T (where T could be S itself) associates with each element x of S an element y of T. We shall write $x \mapsto y$, and if the mapping is called f, we write $f : S \to T$ and one often writes $y = f(x)$ (as for functions), or $y = fx$. This means that when two mappings f, g are performed in succession, the result is $g(f(x))$. This is again a mapping, usually written as gf; thus "first f, then g" is written as gf. This reversal of factors can be avoided by writing mappings on the right: instead of fx one puts xf. With this convention "first f, then g" becomes fg. Naturally these two ways of writing mappings are equivalent and it is important to be familiar with both, since both are needed. We have emphasized this point, because in this book we shall as a rule write mappings on the right: $x \mapsto xf$; as a result the formulae sometimes become a little simpler.

Finally let us recall some basic properties of mappings. Given a mapping between sets, $f : S \to T$, xf is the **image** of x under f, and $Sf = \{xf | x \in S\}$ is the **image** of S; f is called **injective** or **one-one** if distinct elements have distinct images: $xf = yf$ implies $x = y$; f is called **surjective** or **onto** if each element of T is an image: for each $y \in T$ there exists $x \in S$ such that $xf = y$. A mapping that is one-one and onto is called **bijective** or a **bijection**; an older name is "one-one correspondence". The mapping $f : S \to S$, given by $xf = x$, leaving all elements fixed, is called the **identity mapping** on S. Mappings between groups are nearly always assumed to preserve the group structure: $(xy)f = xf.yf$; they are the **homomorphisms**. The image of a homomorphism $f : G \to H$ is a subgroup of H, while the **kernel** of f, consisting of all elements of G mapping to 1, denoted by ker f, is a **normal** subgroup of G, i.e. a subgroup admitting all conjugations $x \mapsto a^{-1}xa$; thus if $x \in \ker f$, then $a^{-1}xa \in \ker f$ for all $a \in G$.

Mappings are often arranged as diagrams. A diagram such as the rectangle of mappings shown below is said to **commute** if the mappings from one point to another along any path give the same result, e.g. in the diagram below, commutativity means that we have $fh = gk$.

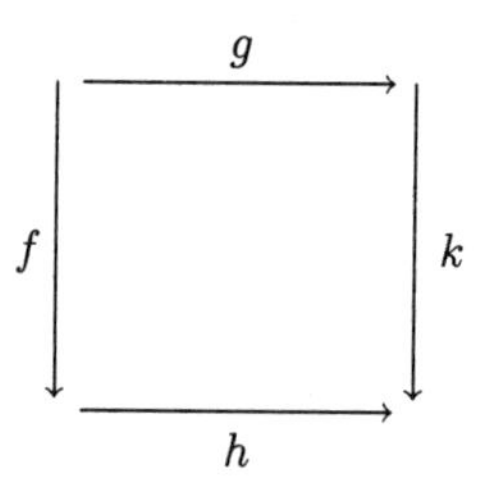

A brief word about proofs may be helpful. Many theorems are in the form of a number of conditions a, b, etc., which are to be proved equivalent. We write (a) $\Rightarrow$ (b) to indicate that (a) implies (b), or that this implication is asserted, and (a) $\Leftrightarrow$ (b) to indicate that (a) and (b) are (to be proved) equivalent, i.e. (a) $\Rightarrow$ (b) and (b) $\Rightarrow$ (a). To prove (a) $\Rightarrow$ (b) one may also prove not(b) $\Rightarrow$ not(a). Sometimes, in order to prove (a) $\Rightarrow$ (b), one assumes (a) and not(b) and derives a contradiction; if we know that (a) holds, it follows that (b) must hold. For any theorem, the end of the proof, or its absence, when it is very easy, is indicated by $\blacksquare$.

It has long been customary to denote ideals by Gothic (Fraktur) letters; however, bearing in mind the struggles of generations of students trying to decipher these symbols, I have decided to use Latin letters. But to give at least a taste of pleasures missed, I have used $\mathfrak{M}$ (Gothic M) for the full matrix ring, though mindful of the problems this may cause, I have included the written form in the list of notations.

In the statement of necessary and sufficient conditions the phrase "if and only if" recurs frequently, and we shall follow the convention that has now become customary in abbreviating it by "iff", except in formal enunciations.

1
Basics

The example of a ring familiar to all of us is the ring of integers. At an early stage we learn to count, as well as add numbers; once we get to school, we meet negative numbers which make subtraction possible in all cases. Later, when we come to study mathematics in earnest, we single out the laws satisfied by integers and consider other systems having two operations satisfying these laws. It turns out that such systems, called **rings**, occur frequently and they form the subject of this volume.

1.1 The Definitions

The following definition captures many of the essential properties of the integers. By a **ring** we understand a system R with two binary operations, associating with each pair of elements a, b a **sum** $a + b$ and a **product** $a.b$ or simply ab, again lying in R and satisfying the following laws:

R.1 $a + b = b + a$ (commutative law of addition);

R.2 $(a + b) + c = a + (b + c)$ (associative law of addition);

R.3 there exists $0 \in R$ such that $a + 0 = a$ (neutral element for addition);

R.4 for each $a \in R$ there exists $-a \in R$ such that $a + (-a) = 0$ (additive inverse);

R.5 $a(bc) = (ab)c$ (associative law of multiplication);

R.6 $(a+b)c = ac + bc$, $c(a+b) = ca + cb$ (distributive laws);

R.7 there exists $1 \in R$ such that $1a = a1 = a$

(neutral element for multiplication).

Readers who have met groups before will immediately see that R.1–R.4 just express the fact that a ring is an abelian (i.e. commutative) group under addition. A system with a single binary operation, which is associative, is called a **semigroup**; from R.5 we see that a ring is a semigroup under multiplication. In fact it is a **monoid**, that is, a semigroup with an element 1 which is a neutral for the multiplication. The neutral for addition in the ring is called the **zero element** or **zero** of R and is written 0, while 1, the neutral for multiplication postulated in R.7 is called the **unit element** or **one** of R. There can only be one such element, for if $1'$ is another one, then we have $1' = 1'1 = 1$, showing $1'$ to be equal to 1. Similarly, the zero 0 is uniquely determined by R.3.

Many rings satisfy the further rule:

R.8 $ab = ba$ for all $a, b \in R$ (commutative law of multiplication).

They are the **commutative** rings; all other rings are **non-commutative**.

Before going on, let us give some examples of rings. Perhaps the most obvious one is the ring of integers: $1, 2, 3\ldots, 0, -1, -2, -3, \ldots$, usually denoted by $\mathbb{Z}$ (which stands for the German for "number": *Zahl*). The laws R.1–R.7 are easily verified in this case. An even simpler system is the set $\mathbb{N} = \{0, 1, 2, 3, \ldots\}$ of all natural numbers; of course this is not a ring, but merely a semigroup, even a monoid, under addition, with neutral element 0, as well as a monoid under multiplication, with neutral element 1. Other examples:

1. The set $\mathbb{Q}$ of rational numbers a/b, where $a, b \in \mathbb{Z}$, $b \neq 0$, and $a/b = a'/b'$ iff $ab' = ba'$, forms a ring; likewise the set $\mathbb{R}$ of all real numbers and the set $\mathbb{C}$ of all complex numbers (see Section 3.6 below).
2. The set $\mathbb{Q}[x]$ of all polynomials in x with rational coefficients is a ring. More generally, the coefficients can be taken from any ring; a formal definition will be given later, in Chapter 3, but we shall occasionally use polynomials with coefficients in a field informally.
3. Let A be an abelian group, written additively: thus we have the operation $(a, b) \mapsto a + b$, where $a + b = b + a$ and $a \mapsto -a$, such that $a + (-a) = 0$. Now consider End(A), the set of all **endomorphisms** of A (i.e. homomorphisms of A into itself). We can define an operation of addition on End(A) by putting $x(\alpha + \beta) = x\alpha + x\beta$, for any $x \in A$ and $\alpha, \beta \in \text{End}(A)$. To verify that this is indeed an endomorphism, we have

$$\begin{aligned}(x+y)(\alpha+\beta) &= (x+y)\alpha+(x+y)\beta\\ &= x\alpha+y\alpha+x\beta+y\beta\\ &= x\alpha+x\beta+y\alpha+y\beta\\ &= x(\alpha+\beta)+y(\alpha+\beta).\end{aligned}$$

We note that this argument depended on the commutativity of A; it breaks down for general groups. Besides addition there is also a multiplication on $\text{End}(A)$, defined by putting $x(\alpha\beta) = (x\alpha)\beta$. Let us show that $\text{End}(A)$ is a ring under these operations. The commutativity and associativity of addition follow from the fact that they hold in A; likewise the zero 0 and unit element 1 are defined by $x0 = 0$, $x1 = x$ for all $x \in A$. To prove that multiplication is associative, we have $x((\alpha\beta)\gamma) = (x(\alpha\beta))\gamma = ((x\alpha)\beta)\gamma = (x\alpha)(\beta\gamma) = x(\alpha(\beta\gamma))$. Finally, to prove the distributive laws, we have $x(\alpha(\beta+\gamma)) = (x\alpha)(\beta+\gamma) = (x\alpha)\beta + (x\alpha)\gamma = x(\alpha\beta) + x(\alpha\gamma) = x(\alpha\beta+\alpha\gamma)$; the other distributive law is verified in just the same way.

We also note that the ring obtained in this way is not usually commutative. For example, let A be the group of four elements, $0, a, b, c$ with a, b, c of order two and $a + b = c$ (the **Klein four-group**, after Felix Klein, 1849–1925). A has an automorphism α interchanging a and b (and leaving c fixed) and an automorphism β interchanging a and c (and leaving b fixed). Then $\alpha\beta$ and $\beta\alpha$ each permute a, b, c in a 3-cycle, but are different; in fact one is the inverse of the other, as is easily checked.

4. Let I be any set and denote by $\mathbb{Z}^I$ the set of all functions from I to $\mathbb{Z}$; thus the elements of $\mathbb{Z}^I$ are families (x_i) of integers indexed by I. Addition and multiplication are defined componentwise: $(x_i) + (y_i) = (z_i)$, where $z_i = x_i + y_i$ and $(x_i)(y_i) = (t_i)$, where $t_i = x_i y_i$. It is easily verified that with these rules $\mathbb{Z}^I$ becomes a ring, the **function ring** on I with values in $\mathbb{Z}$.
5. A ring consisting of finitely many elements, $\{a_1, \ldots, a_n\}$ can be described by giving the n^2 sums $a_i + a_j$ in a table, the **addition table** and the n^2 products $a_i a_j$ in the **multiplication table**. For example let us form a ring on two elements. The two elements must be 0 and 1 and the multiplication table is completely determined, because all products are given by R.3 and R.7. Of the sums, only $1 + 1$ remains to be determined, and this cannot be 1, because the group property would then allow us to cancel 1, giving $1 = 0$. Hence $1 + 1 = 0$ and the tables take the form

+	0	1
0	0	1
1	1	0

×	**0**	**1**
0	**0**	**0**
1	**0**	**1**

6. An even simpler example is the ring consisting of a single element; this must be 0 and moreover $1 = 0$. This is the **zero ring** or **trivial** ring, denoted by $\{0\}$ or also 0. We shall usually not be interested in this ring, as it only occurs as a limiting case.
7. The set of even integers with the usual addition and multiplication forms a system satisfying R.1–R.6; thus it is a ring except that it lacks a unit-element. Sometimes we may need to consider such "rings without one", which may be called **rungs** (to have a name). We remark that in some books rings are not required to have a unit element, but since most systems used in practice actually do have a one, it is more practical to postulate its existence from the outset.
8. Given any ring R, we can form another ring from it by taking the additive group of R and defining multiplication by the rule: $x.y = yx$; in other words, we multiply elements in the opposite order. This ring is called the **opposite** of R and is denoted by R°. For a commutative ring R we have $R^\circ = R$; later we shall meet examples of non-commutative rings that are isomorphic to their opposite, but equality can hold only when the ring is commutative. Further we note that $R^{\circ\circ} \cong R$ for any ring R.

In most of the above examples the ring is commutative. The ring $\mathbb{Q}$ of rational numbers has another important property: it allows division. A commutative ring, not the zero ring, in which the set of all non-zero elements is a group under multiplication is called a **field**; thus the rational numbers $\mathbb{Q}$, the real numbers $\mathbb{R}$, and the complex numbers $\mathbb{C}$ are examples of fields. The complex numbers have the further property of being **algebraically closed**; this means that every polynomial equation with complex coefficients has a complex root, and it is in contrast to $\mathbb{Q}$, where $x^2 = 2$ has no root, and $\mathbb{R}$, where $x^2 = -1$ has no root. Example 5 above is also a field; in Section 1.2 below we shall meet other finite fields. Let us define, quite generally, a **division ring** or **skew field** as a non-zero ring R (i.e. not the zero ring) in which the non-zero elements form a group under multiplication; later (in Sections 2.1 and 5.2) we shall meet some non-commutative examples.

As in the case of groups, we have the notions of homomorphism, isomorphism, endomorphism and automorphism. By a **homomorphism** of a ring

R into a ring S we understand a mapping $f : R \to S$ which preserves all the ring operations, i.e.

$$(x + y)f = xf + yf, (xy)f = xf.yf, \quad \text{for all} \quad x, y \in R, \quad 1_R.f = 1_S. \quad (1.1)$$

Here we have denoted the ones of R and S by 1_R, 1_S, but in future we shall usually omit the suffix. We remark that the last equation in (1.1) does not follow from the others, and so has to be postulated separately; on the other hand we have $0_R f = 0_S$ as a consequence of the fact that every ring is a group under addition. As an example, the mapping f given by $xf = 0$ for all $x \in R$, satisfies the first two laws of (1.1) but not the third, unless S is the zero ring.

An **isomorphism** $f : R \to S$ is defined as a homomorphism f which has an inverse $g : S \to R$; thus fg is the identity on R and gf the identity on S. We shall write $R \cong S$ to indicate that R is isomorphic to S. If both R and S are extensions of a ring A, then any isomorphism between R and S which reduces to the identity mapping on A is called an isomorphism **over** A. A homomorphism from R to itself is called an **endomorphism** and an endomorphism that is also an isomorphism is called an **automorphism**. This is in agreement with the usage for other systems such as groups, etc. It is clear that an isomorphism between rings R and S is surjective and injective, i.e. **bijective**; conversely, any homomorphism that is bijective is an isomorphism, for it then has an inverse and the latter is a homomorphism by (1.1): given $(xy)f = xf.yf$, let us write $xf = u$, $yf = v$ and apply f^{-1}. Then we obtain $xy = (uv)f^{-1}$, i.e. $(uf^{-1})(vf^{-1}) = (uv)f^{-1}$, and similarly for the other laws.

Let R be any ring. By a **subring** of R we understand a subset R' which is a ring under the operations of R. This means that R' admits addition, multiplication and it contains the unit element of R. For example, the integers $\mathbb{Z}$ form a subring of $\mathbb{Q}$, the ring of rational numbers. The set E of even integers forms a subset of $\mathbb{Z}$ satisfying R.1–R.6 but not R.7 because the unit element 1 of $\mathbb{Z}$ is odd and so is not in E. As another example, the image of any ring homomorphism from R to S is a subring of S, as is easily verified. An injective homomorphism is also called an **embedding** and R is said to be **embedded** in S, and S is an **extension ring** of R, if there is an injective homomorphism from R to S. Of course this does not mean that R is a subring of S, but merely that R is isomorphic to a subring R'' of S. Nevertheless, we can find a ring S' isomorphic to S of which R is a subring and such that the isomorphism from S' to S maps R to R''. We simply take an isomorphic copy S' of S and in it replace each element of S' corresponding to an element of R'' by the corresponding element of R.

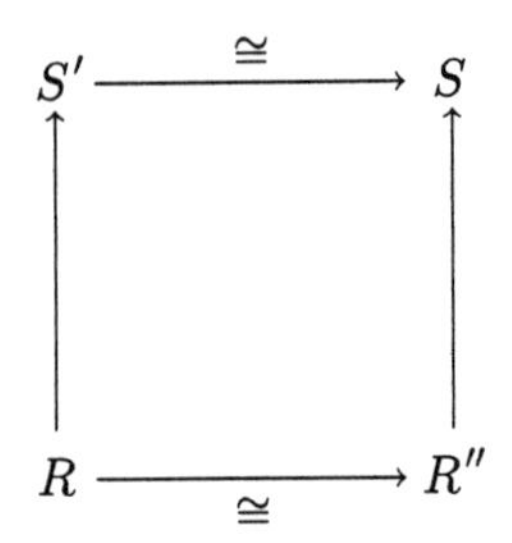

If $f : R \to S$ is an isomorphism of additive groups reversing the multiplication: $(xy)f = yf.xf$, then f is called an **anti-isomorphism**; thus every ring is anti-isomorphic to its opposite. Similarly, an **anti-automorphism** of R is an anti-isomorphism of R with itself; an anti-automorphism (not the identity) whose square is the identity is called an **involution**.

If R is a ring and $\{A_\lambda\}$ is a family of subrings of R, then the intersection $T = \bigcap A_\lambda$ is again a subring, for any ring operation carried out on elements in T gives a result lying in each A_λ and hence in T. Given any ring R and a subset X of R, we can consider the set of all subrings of R containing X and form their intersection, T say. By what has been said, this is a subring of R, clearly the smallest subring containing X. Another way of forming T is to take the set T' of all finite sums of products of elements of $X : \sum_i a_{i1}a_{i2}\dots a_{in_i}$, $a_{ij} \in X$. This is a subring, if we remember that in forming sums of products we must include the empty sum, which is 0 and the empty product, which is 1. Clearly $T' \supseteq X$, and so $T' \supseteq T$; but any element in T' is a sum of products of elements of X, and so lies in each subring containing X, and therefore lies in T. This then proves that $T' = T$. This ring T is called the subring of R **generated** by X and X is called a **generating set** of T; we can sum up our findings as

Theorem 1.1

Given a ring R and a subset X of R, the subring of R generated by X may be obtained as the intersection of all subrings of R containing X. ■

To give a simple example, consider the ring of all complex numbers of the form $a + b\mathrm{i}$, where $a, b \in \mathbb{Z}$ and $\mathrm{i} = \sqrt{-1}$; this is generated by i, since any complex number in our example has the form $a + b\mathrm{i}$, where $a, b \in \mathbb{Z}$.

Besides subrings, another class of subsets of rings play an important role: the ideals. An **ideal** is defined as a subgroup A of R such that $ar, ra \in A$ for all $a \in A$, $r \in R$. One often writes $A \lhd R$ to indicate that A is an ideal of R. For example, in the ring $\mathbb{Z}$ of integers the set of all multiples of any fixed number is an ideal; in a moment we shall see that there are no other ideals in $\mathbb{Z}$, which makes this ring rather special. In every ring the subset consisting of 0 forms an ideal, the **zero ideal**, also called the **trivial** ideal.

From the definition we see that an ideal is always closed under addition, subtraction and multiplication, but it does not in general contain 1, and so need not be a subring of R. In fact if an ideal A of R contains 1, then it must contain $x = x.1$, for any $x \in R$ and so $A = R$. An ideal is said to be **proper** if it is different from R.

Ideals play an important role in analysing homomorphisms of rings. To give an example, let us take the ring of integers $\mathbb{Z}$. To begin with let us find all the ideals in $\mathbb{Z}$. Fix a positive integer m; then the set $m\mathbb{Z} = \{ma \,|\, a \in \mathbb{Z}\}$ is an ideal in $\mathbb{Z}$. Moreover, every ideal A is of this form. For either $A = \{0\}$, the zero ideal, or A contains non-zero integers; with any number n, $-n$ also lies in A, so A contains positive integers. Let m be the least positive integer in A; then for any $n \in A$ we have, by the division with remainder, $n = qm + r$, where $0 \leq r < m$. Hence $r = n - qm \in A$, but m was the least positive integer in A, therefore $r = 0$ and so $n = qm$. Thus $A = m\mathbb{Z}$; an ideal like $m\mathbb{Z}$, also written (m), consisting of all multiples of a single element, is called a **principal** ideal; so we see that every ideal in $\mathbb{Z}$ is principal.

Now for every positive integer m we can define a ring of m elements, denoted by $\mathbb{Z}/(m)$, by taking the numbers $0, 1, 2, \ldots, m-1$ and performing the usual addition and multiplication, followed by taking the residue mod m, where m is called the **modulus**; "mod" stands for **modulo** and means "up to multiples of". This is also called "addition and multiplication mod m"; we say a is **congruent to** b mod m and write $a \equiv b \pmod{m}$, an "equality mod m", which is called a **congruence**, to indicate that a differs from b by a multiple of m. The elements congruent to a given integer a form a **residue class** or a **congruence class**. It is easy to verify that the residue classes again form a ring; moreover the mapping from $\mathbb{Z}$ to $\mathbb{Z}/(m)$ which takes each integer to its residue class is a homomorphism. Congruences were first defined by Carl Friedrich Gauss (1777–1855).

As an example let us take the ring $\mathbb{Z}/(2)$. There are two residue classes, the even integers and the odd integers. Any integer n satisfies $n \equiv 0 \pmod{2}$ if n is even and $n \equiv 1 \pmod{2}$ if n is odd, and we can add and multiply the residue classes by taking representatives, e.g. 0 and 1, adding and multiplying in the usual way and then taking the residue class containing the resulting number. In this way we obtain the ring of two elements listed in Example 5 above. Another example is the ring $\mathbb{Z}/(7)$; this ring is used in everyday life to work out on which day of the week a given date falls (although the ring properties are not used, merely the fact that it is an additive group).

In general, ring homomorphisms are very much like this one, though the details may be more complicated, e.g. when the rings are non-commutative. Thus let R, S be any rings and $f : R \to S$ a homomorphism. Then as a homomorphism of the additive groups f has a **kernel**, ker $f = \{x \in R | xf = 0\}$, which is a subgroup of the additive group of R. Further, if $xf = 0$ and a is any element of R, then $(xa)f = xf.af = 0$, $(ax)f = af.xf = 0$, hence ker f is an ideal of R. Conversely we can show that any ideal occurs as the kernel of a ring homomorphism. Thus let A be an ideal of R; then A is an additive subgroup and we can form the **quotient** R/A, whose elements $a + A$ are the **cosets**, i.e. residue classes of the ideal A in R, and the **natural mapping** $\nu : R \to R/A$ taking each element to its coset mod A is a homomorphism of additive groups. Now it only remains to define a ring structure on R/A such that ν becomes a ring homomorphism. Let α, $\beta \in R/A$; to define $\alpha\beta$, we choose $a \in \alpha$, $b \in \beta$ and put $\alpha\beta = ab + A$. If instead of a, b we choose $a' = a + u$, $b' = b + v$, where $u, v \in A$, then $a'b' = ab + av + ub + uv \in ab + A$, and this shows $\alpha\beta$ to depend only on α, β and not on the representatives chosen. It is clear that with this definition the natural mapping $a \mapsto a\nu$ is a homomorphism, and by applying ν to the laws R.5–R.7 we see that R/A with the above definitions is a ring. We can also obtain an analysis of ring homomorphisms in this way. The result may be stated as follows.

Theorem 1.2

Let R be a ring and $f : R \to S$ a homomorphism. Then im *f*, the image of *f*, is a subring of *S*, ker *f*, the kernel of *f*, is an ideal of R and f induces an isomorphism

$$R/\ker f \cong \operatorname{im} f. \tag{1.2}$$

Conversely, given a ring R and an ideal A of R, the additive quotient group R/A can be defined as a ring in such a way that the natural mapping $\nu : R \to R/A$ is a ring homomorphism.

Proof

The assertions about im f and ker f as well as the converse have already been proved. By definition of im f, f maps R onto im f, and two elements a, b of R have the same image iff $af = bf$, i.e. $a - b \in \ker f$. Thus there is a bijection between the residue classes of ker f in R and the elements of im f, which by its definition preserves the ring operations. Hence we have obtained an isomorphism (1.2). ■

Taking again the ring of integers, $\mathbb{Z}$, as an example, as we have seen, we can for any positive integer m form the ideal (m) consisting of all multiples of m. The residue class ring $\mathbb{Z}/(m)$ consists of the m residues modulo m. As a representative for any integer a we may take its least residue after division by $m : a = qm + r$, where $0 \leq r < m$. For example, by mapping even integers to 0 and odd integers to 1, we obtain a homomorphism from $\mathbb{Z}$ to $\mathbb{Z}/(2)$. It is easy to verify that when m is a prime number, (and only then) $\mathbb{Z}/(m)$ is a field. As in the case of $\mathbb{Z}$, for an ideal A in a general ring R one also writes $x \equiv y (\text{mod } A)$ to indicate that $x - y \in A$.

In any ring R, if a subset X of R is given, then there is a smallest ideal containing X. We form the set A of all finite sums $a_1x_1b_1 + a_2x_2b_2 + \ldots + a_rx_rb_r$, where $x_i \in X$ and $a_i, b_i \in R$. It is clear that any ideal containing X must also contain A; but A itself is easily seen to be an ideal, so it is the ideal we are looking for. We shall also say that the ideal A is **generated** by X; e.g. if X is finite, then A is a finitely generated ideal. In the special case when R is commutative, the ideal generated by X consists of all finite sums $\sum a_i x_i$, where $a_i \in R$, $x_i \in X$. In a commutative ring the ideal generated by $x_1, \ldots, x_n$ is often denoted by $(x_1, \ldots, x_n)$.

In any ring R the subset $\mathscr{Z}(R) = \{a \in R | ax = xa \text{ for all } x \in R\}$ is called the **centre** of R and an element of the centre is called **central**; clearly the centre is a commutative subring, for if $a, b \in \mathscr{Z}(R)$, then $ab = ba$ and for any $x \in R$, $(a - b)x = ax - bx = xa - xb = x(a - b)$, $abx = axb = xab$. If R is a skew field, we can even assert that $\mathscr{Z}(R)$ is a field, for we have seen that it is a subring and if $0 \neq a \in \mathscr{Z}(R)$, then for any $x \in R$, $ax = xa$, and on multiplying by a^{-1} on both sides, we find that $xa^{-1} = a^{-1}x$. We state the result as

Theorem 1.3

Let R be any ring. Then its centre $\mathscr{Z}(R)$ is a commutative subring, which is a subfield when R is a skew field. ■

More generally we can, for any subset X of R, define its **centralizer** as $\mathscr{C}_R(X) = \{a \in R | ax = xa \text{ for all } x \in X\}$; any element of $\mathscr{C}_R(X)$ is said to **centralize** X. It is easily seen that $\mathscr{C}_R(X)$ is always a subring of R.

Exercises 1.1

1. Show that R.1 is a consequence of the other ring axioms.
2. If A is a system satisfying R.1–R.6 but not R.7 (a "ring without a 1"), show that the direct sum $A \oplus Z$ can be defined as a ring

by putting $(a,m)+(b,n)=(a+b, \quad m+n)$, $(a,m)(b,n)=(ab+mb+an,mn)$. What is the unit element of this ring?

3. In any ring R, show that for fixed $a \in R$, the mapping $x \mapsto ax$ $(x \in R)$ is always an endomorphism of the additive group of R. What is the condition on a for it to be a ring homomorphism?

4. Write out multiplication tables for $\mathbb{Z}/(3)$ and $\mathbb{Z}/(4)$. Which of these is a field?

5. Consider the ring $\mathbb{Z}^{\mathbb{N}}$, formed as in Example 4 of the text (with $I=\mathbb{N}$). Show that the mapping $(x_1, x_2, \ldots) \mapsto x_1$ is a homomorphism from $\mathbb{Z}^{\mathbb{N}}$ to $\mathbb{Z}$; similarly if the suffix 1 is replaced by any other number.

6. Show that any cyclic group (written additively) can be defined as a ring R by taking the generator u to be the unit element and defining the multiplication in R by the rule $mu.nu = mnu$.

7. Show that in any ring R, the sum of two ideals A and B, defined as $A+B=\{a+b \mid a \in A, b \in B\}$ is again an ideal, as is the product $AB=\{\sum a_i b_i \mid a_i \in A, b_i \in B\}$. If further, $A+B=R$, show that $AB=A \cap B$.

8. If R is any ring and X a subset of R, verify that the intersection of all ideals containing X is the ideal generated by X.

9. Show that every subgroup of the additive group of $\mathbb{Z}$ is an ideal. Which other rings have this property?

10. Verify that the centralizer $\mathscr{C}_R(X)$ of $X \subseteq R$ is a subring of R. Show further that $\mathscr{C}_R(\mathscr{C}_R(X)) \supseteq X$ for all $X \subseteq R$, and that $\mathscr{C}_R(X) \supseteq X$ iff X is commutative.

1.2 Fields and Vector Spaces

In Section 1.1 fields were defined as systems with four operations: addition, subtraction, multiplication and division, satisfying the familiar laws. Put differently, a field F is an abelian group under an operation $+$ with neutral 0, such that $F \setminus \{0\}$ is an abelian group under a second operation $\times$, and these two operations are linked by the distributive laws. The group $F \setminus \{0\}$ is denoted by $F^\times$ and is called the **multiplicative group** of F. Of course there is no need to assume $F^\times$ commutative; in the general case (as already mentioned in Section 1.1) F is called a **skew field** or **division ring**.

In any ring R a **unit** or **invertible element** is an element $a \in R$ with a two-sided inverse; thus there exists $a' \in R$ such that $aa' = a'a = 1$. The inverse is uniquely determined by a, for if a'' is another inverse, then we have $a' = a'.1 = a'aa'' = 1.a'' = a''$, where we have used the associative law. The inverse of a is usually denoted by a^{-1}; in a commutative ring one could also write it as $1/a$, but this would be ambiguous in a general ring. Clearly a non-zero ring is a skew field precisely when every non-zero element is a unit.

To study fields we shall need another property. A ring R is called an **integral domain**, if $1 \neq 0$ in R and for any elements $a, b \in R, ab = 0$ implies that $a = 0$ or $b = 0$. As a consequence we have the following **cancellation rule** in any integral domain.

Proposition 1.4

In an integral domain R, if $ac = bc$ or $ca = cb$, where $c \neq 0$, then $a = b$.

For when $ac = bc$, then $(a - b)c = 0$, and since $c \neq 0$, it follows that $a - b = 0$, so $a = b$. Similarly, if $ca = cb$ and $c \neq 0$, then $a = b$. ■

This rule is of course familiar from the integers, but we must bear in mind that it no longer holds in general rings.

For an integral domain R we shall write $R^\times$ for the set of all non-zero elements of R (we note that this agrees with the usage introduced earlier for fields); clearly $R^\times$ is a monoid under ring multiplication. Further, it is easy to see that any field and, more generally, any subring of a field is an integral domain, for $1 \neq 0$ by definition and if $ab = 0$ and $a \neq 0$, say, then $b = a^{-1}ab = 0$. However, not every integral domain is a field, as is shown by the example of the ring $\mathbb{Z}$ of integers; but there is one special case where the converse holds:

Theorem 1.5

Every finite commutative integral domain is a field.

Proof

Let $R = \{a_0, a_1, \ldots, a_n\}$ be a commutative integral domain. Then for any non-zero element c of R the elements $a_0c, \ldots, a_nc$ are all different, because $a_ic - a_jc = (a_i - a_j)c \neq 0$ for $i \neq j$. Hence they again form a set of $n + 1$ elements, so every element of R occurs among them. Thus the equation $xc = b$ has a solution for all $b, c \in R$, where $c \neq 0$. In particular, c has an inverse

c^{-1}, obtained by solving $xc = 1$, and it follows that the non-zero elements of R form a group under multiplication, as we wished to show. ■

The observant reader will have noticed that the commutativity of R was not really used, so that the same proof will show that a "non-commutative integral domain" which is finite must be a skew field. But now one can invoke a more advanced theorem which states that a skew field that is finite must be commutative; this result goes beyond the framework of this book (see, e.g. A.2, p. 101) and so is not included here. For this reason there is no point in proving Theorem 1.5 in the skew case.

An important source of fields is given by forming the field of fractions of an integral domain, just as the rational numbers were formed as fractions of integers. It turns out that this method applies quite generally:

Theorem 1.6

Let R be a commutative integral domain. Then there exists a field F containing R as a subring and generated by it as a field. This field F is determined up to isomorphism by R and its elements have the form a/b, where $a, b \in R, b \neq 0$ and $a/b = a'/b'$ if and only if $ab' = ba'$.

The proof consists in defining an equivalence relation on the set $R \times R^{\times}$ by putting

$$(a, b) \sim (a', b') \text{ iff } ab' = ba', \tag{1.3}$$

and verifying that these classes form a field containing a copy of R. To show that (1.3) is an equivalence, clearly it is reflexive: $(a, b) \sim (a, b)$; it is also symmetric: if $(a, b) \sim (a', b')$, then $ab' = ba'$, hence $(a', b') \sim (a, b)$; finally to prove transitivity, if $(a, b) \sim (a', b')$ and $(a', b') \sim (a'', b'')$, then $ab' = ba'$, $a'b'' = b'a''$, hence $ab'b'' = ba'b'' = bb'a''$ and cancelling $b'(\neq 0)$, we obtain $ab'' = ba''$, therefore $(a, b) \sim (a'', b'')$. Let us denote the equivalence class of (a, b) by a/b and the set of all a/b by F, so that $a/b = a'/b'$ precisely when $ab' = ba'$. We can define addition and multiplication on F by the equations

$$a/b + c/d = (ad + bc)/bd, \qquad a/b.c/d = ac/bd. \tag{1.4}$$

Here we must show that the element on the right-hand side depends only on the equivalence classes of the elements on the left, not on a, b, c, d themselves; this is straightforward and so may be left to the reader. Secondly the ring laws must be checked for the sum and product defined in (1.4); this is also a routine verification. The elements $a/1$ ($a \in R$) form a subring isomorphic to R, and so R may be embedded as a subring in F. Further, F is a field since $a/b \neq 0$ whenever $a \neq 0$ and then $(a/b)^{-1} = b/a$; it is generated by R because $a/b = (a/1)(b/1)^{-1}$. If G is another field generated by a subring isomorphic to R, then if f is the isomorphism from R to a subring of G, we have $(a/b)f = (a/1)f.[(b/1)f]^{-1}$; thus f extends to an isomorphism from F to G over R, i.e. leaving R fixed. ■

The field constructed in Theorem 1.6 is called the **field of fractions** of R. The following result is a slight generalization; here a **regular** element is an element c such that $cx = 0$ implies $x = 0$ and a set T of elements of a ring is **multiplicative** if $1 \in T$ and $a, b \in T$ implies $ab \in T$.

Theorem 1.7

Let R be any commutative ring and T a multiplicative subset of regular elements of R. Then there is a ring R_T containing R as subring as well as inverses of all the elements of T, and generated by R and these inverses. Moreover, R_T is determined up to isomorphism by R and T.

The proof is as for Theorem 1.6, and so will not be repeated. The ring R_T is called the **localization** of R at T. ■

There is a useful generalization of Theorems 1.6 and 1.7 to certain types of non-commutative rings (the Ore domains, see Section 5.1), but the analogue of Theorem 1.6 for general integral domains does not hold (see Section 4.7).

Let R be any integral domain. Given $a \in R$, we can form multiples of $a : a,\ a + a,\ a + a + a, \ldots$, which can briefly be written as $a,\ 2a,\ 3a, \ldots$, but care is needed because it may happen that $na = 0$ even though $n \neq 0$. Excluding the trivial case when $a = 0$, if $na = 0$ and $a \neq 0$, then we can cancel a, so $n.1 = 0$, and hence $nx = 0$ for all $x \in R$, so it is enough to consider the multiples of 1. The least multiple of 1 equal to 0 is just the order of 1, as member of the additive group of R. This is called the **characteristic** of R; if $n.1 \neq 0$ for all $n \neq 0$, R is said to have characteristic zero, although the term "infinite characteristic" is also sometimes used. If 1 has finite order n, this order must be a prime; for $1.1 = 1 \neq 0$, so $n \neq 1$, and if $n = rs$, then $(r.1)(s.1) = rs.1 = 0$, and so either $r.1 = 0$ or $s.1 = 0$. Thus the least n such that $n.1 = 0$ is a prime number. All this applies in particular to any field, even skew. Thus we have proved

Theorem 1.8

The characteristic of any integral domain is either zero or a prime number. In particular, this holds for any field (even skew). ■

In any field (even a skew field) F we can form the subfield F_0 generated by 1; strictly speaking we can say that F_0 is generated by the empty set, for every subfield of F must contain the 1 of F. Hence F_0 is the least subfield, called the **prime subfield** of F, and the mapping

$$f : n \mapsto n.1 \tag{1.5}$$

is a homomorphism of $\mathbb{Z}$ into F. If F has characteristic 0, this homomorphism is an embedding and it can then be extended to a homomorphism of $\mathbb{Q}$ into F, by the rule: $(ab^{-1})f = af(bf)^{-1}$; writing a/b for ab^{-1}, we have $(a/b)f = af/bf$. Here some care is needed, for there is more than one way of expressing an element of $\mathbb{Q}$ as a fraction: if $a/b = a'/b'$, all we can say is that $ab' = ba'$. But it then follows that $(ab')f = (ba')f$, i.e. $af.b'f = bf.a'f$ and so $af/bf = a'f/b'f$. This shows that the definition of f for an element u of $\mathbb{Q}$ is independent of the way of representing u, so that f is well defined on the whole of $\mathbb{Q}$. Here we used the fact that a, b, a', b' all lie in a commutative ring, namely $\mathbb{Z}$. Suppose now that F has prime characteristic p. Then the mapping (1.5) has the kernel (p) in $\mathbb{Z}$ and by Theorem 1.2, $F_0 \cong \mathbb{Z}/(p)$; thus the prime subfield is the field of p elements. This field is also denoted by $\mathbb{F}_p$; thus we can express the result as

Theorem 1.9

In any field (even skew) F the subfield generated by 1 is the prime subfield. It is isomorphic to $\mathbb{Q}$ if F has characteristic zero and to $\mathbb{F}_p$ if F has prime characteristic p. ■

No doubt all readers will have met vector spaces before. We recall that a **vector space** over a field F, briefly an F-**space**, is an abelian group V on which the elements of F operate by endomorphisms, subject to the rules:

V.1 $\alpha(x+y) = \alpha x + \alpha y$ for $\alpha \in F$, $x, y \in V$;

V.2 $(\alpha+\beta)x = \alpha x + \beta x$ for $\alpha, \beta \in F$, $x \in V$;

V.3 $(\alpha\beta)x = \alpha(\beta x)$ for $\alpha, \beta \in F$, $x \in V$;

V.4 $1.x = x$ for all $x \in V$.

The rule V.1 just tells us that each α defines an endomorphism of V, so that we have a mapping from F to $\mathrm{End}(V)$, and now V.2–V.4 express the fact that this mapping is a ring homomorphism. Here we have not had to use the commutativity of F, so that all that is said applies to skew fields, but we shall simply talk about fields, since that is the case of main interest to us. Vector spaces over a field have a particularly simple structure, which is obtained by using the field properties of F. If V is any vector space, then a **subspace** of V is defined as a non-empty subset V' of V which is closed under addition and multiplication by elements of F. Given any subset X of a vector space V, we can form the set of all linear combinations of $X : \alpha_1 x_1 + \ldots + \alpha_r x_r$ where $\alpha_i \in F$, $x_i \in X$ for $i = 1, \ldots, r$. This is easily verified to be a subspace; it is the subspace **spanned** by X. Thus if we assume that V is spanned by a finite set, then we know that V has a basis $u_1, \ldots, u_n$ which has the property that every element of V can be written in the form

$$\xi_1 u_1 + \ldots + \xi_n u_n, \qquad (\xi_i \in F) \tag{1.6}$$

where the coefficients ξ_i are uniquely determined (cf. A.1, Chapter 5). Equivalently, the u_i form a basis if every vector of V can be written in the form (1.6) and the u_i are **linearly independent**, i.e. an expression of the form (1.6) vanishes iff all the ξ_i vanish. An expression such as (1.6), i.e. a linear combination of the u's, will often be abbreviated by $\sum_i \xi_i u_i$ (using the Greek letter sigma, $\sum$, for "sum"), or simply $\sum \xi_i u_i$, where the range of summation can be gathered from the context. It is useful to remember that as a rough guide, the summation of such formal expressions is usually over a repeated suffix. This is true so often that in physics the $\sum$-sign in a summation is often omitted, with the convention that the summation is over each repeated index. We shall not follow this convention, but always write out the $\sum$-signs.

Without the assumption of finite generation it is still true that each vector space has a basis, but this may now be infinite and the proofs are rather harder. Usually one then makes other assumptions about the space, and this is really a separate topic, so we shall here confine ourselves to vector spaces that are finitely generated. In A.1, Chapter 5, it is shown that any such space has a basis and all bases of a given space V have the same number of elements, called the **dimension** of V and written $\dim V$. As we shall see below, the existence of a basis greatly simplifies the treatment of vector spaces. Thus if $u_1, \ldots, u_n$ is a basis of V and the subspace spanned by u_i is denoted by U_i, then each U_i is one-dimensional and the fact that the u_i span V is expressed by saying that V is the sum of its subspaces $U_1, \ldots, U_n$. Generally a vector space V is said to be the sum of its subspaces $V_1, \ldots, V_r$ and we write $V = V_1 + \ldots + V_r$ if every vector in V can be written as a sum of vectors in the V_i: thus any $v \in V$ is of the form

$$v = v_1 + \ldots + v_r, \qquad v_i \in V_i.$$

If this expression is unique for each $v \in V$, the sum is said to be **direct** and it is written

$$V = V_1 \oplus \ldots \oplus V_r.$$

Thus in the example given earlier, V is the direct sum of its one-dimensional subspaces U_i.

We have already seen that the endomorphisms of an abelian group are of importance; when we have a vector space, we shall be interested in endomorphisms that are compatible with the action of the field. Thus for a vector space V over a field F we define a **linear transformation** or **linear mapping** as a mapping $f : V \to V$ such that

L.1 $(x + y)f = xf + yf$, for all $x, y \in V$;

L.2 $(\alpha x)f = \alpha(xf)$ for all $x \in V, \alpha \in F$.

We see from L.1 that a linear transformation is an endomorphism, while L.2 states that it commutes with the action of the field. If we think of the field acting on V as a subring of End(V), we see that the linear transformations are just the endomorphisms commuting with all the field operations; but this commutative law is expressed as an associative law in L.2 because we have written the field operations on the left and the linear transformations on the right of the elements of V. The set of all linear transformations of V over F is denoted by $\mathscr{L}_F(V)$. This is a subring of End(V), whose members can be written in a particularly simple form, once a basis of the space has been chosen. Thus let $u_1, \ldots, u_n$ be a basis of V; any element of V can be written in the form (1.6), so we have in particular, $u_i f = \sum_j \alpha_{ij} u_j$, where $\alpha_{ij} \in F$. If the general element of V is $x = \sum_i \xi_i u_i$, then $xf = \sum(\xi_i u_i)f = \sum \xi_i(u_i f) = \sum \xi_i(\sum \alpha_{ij} u_j) = \sum \xi'_j u_j$, where $\xi'_j = \sum \xi_i \alpha_{ij}$. The set of n^2 elements (α_{ij}) is called the **matrix** representing f in the basis (u_i). We note that (i) the matrix (α_{ij}) determines the action of f completely, and (ii) any choice of n^2 elements α_{ij} of F defines a linear transformation of V. The matrix (α_{ij}) is usually written as an $n \times n$ square, where the first suffix indicates the row and the second the column, e.g. for $n = 2$, the matrix (a_{ij}), written out in full, is

$$\begin{pmatrix} a_{11} & a_{12} \\ a_{21} & a_{22} \end{pmatrix}.$$

Such a matrix is often denoted by a single letter, A say, and the set of all $n \times n$ (read: n by n) matrices is called the **full** $n \times n$ **matrix ring** over F, written $\mathfrak{M}_n(F)$ or sometimes simply F_n. As we have seen, the choice of a basis in V establishes a bijection between $\mathscr{L}_F(V)$ and $\mathfrak{M}_n(F)$. This bijection will be an isomorphism, provided that we define the addition and multiplication of matrices by the rules:

$$(\alpha_{ij}) + (\beta_{ij}) = (\alpha_{ij} + \beta_{ij}), \quad (\alpha_{ij})(\beta_{ij}) = (\gamma_{ij}), \quad \text{where } \gamma_{ij} = \sum_r \alpha_{ir}\beta_{rj}. \tag{1.7}$$

These are of course the familiar laws of matrix addition and multiplication; they have been chosen precisely in order to have this isomorphism between $\mathscr{L}_F(V)$ and $\mathfrak{M}_n(F)$. The addition is straightforward: we add componentwise. For the multiplication we form the (i, j)-entry of the product, i.e. the entry in the i-th row and j-th column, by multiplying the i-th row of the first factor into the j-th column of the second factor, i.e. multiplying corresponding terms and forming their sum. To define the unit matrix, we shall need a function called the **Kronecker delta** (after Leopold Kronecker, 1823–1891), which is a function of two variables, defined as

$$\delta_{ij} = \begin{cases} 1 & \text{if } i = j, \\ 0 & \text{if } i \neq j. \end{cases}$$

Now the unit-element of $\mathfrak{M}_n(F)$ is the **unit matrix** I, defined by $I = (\delta_{ij})$. It has ones along the diagonal $i = j$, called the **main diagonal**, and zeros elsewhere.

Of course it is important to bear in mind that the isomorphism $\mathscr{L}(V) \cong \mathfrak{M}_n(F)$ depends on the choice of basis in V. Let us see how it changes when the basis is changed. If $v_1, \dots, v_n$ is a second basis of V, we know from A.1, Chapter 5, or also Theorem 1.25 below, that it also has n elements, and the transformation from the u's to the v's has the form $v_i = \sum \pi_{ij} u_j$; if $P = (\pi_{ij})$, then it is convenient to write the basis (v_i) as a column, $\boldsymbol{v}$ say, and likewise write (u_i) as a column $\boldsymbol{u}$. Now the transformation from $\boldsymbol{u}$ to $\boldsymbol{v}$ can be written as $\boldsymbol{v} = P\boldsymbol{u}$, where we think of $\boldsymbol{u}$, $\boldsymbol{v}$ as matrices with a single column and form the matrix product of P and $\boldsymbol{u}$ to obtain $\boldsymbol{v}$. Clearly the transformation from $\boldsymbol{v}$ to $\boldsymbol{u}$ is $\boldsymbol{u} = P^{-1}\boldsymbol{v}$, using the inverse of P. If our linear transformation is $\boldsymbol{u}f = A\boldsymbol{u}$, or in terms of $\boldsymbol{v}$, $\boldsymbol{v}f = B\boldsymbol{v}$, then on replacing $\boldsymbol{v}$ by its value $P\boldsymbol{u}$, we obtain $(P\boldsymbol{u})f = B(P\boldsymbol{u})$, which by linearity becomes $P(\boldsymbol{u}f) = BP\boldsymbol{u}$, i.e. $\boldsymbol{u}f = P^{-1}BP\boldsymbol{u}$. Since $\boldsymbol{u}f = A\boldsymbol{u}$, we obtain

$$A = P^{-1}BP. \tag{1.8}$$

Two matrices A and B related as in (1.8) are said to be **similar**; so a given linear transformation is represented in different bases by matrices that are similar. Moreover, if f is represented by A in one basis, then any matrix similar to A represents f in another suitably chosen basis. For suppose that $\boldsymbol{u}f = A\boldsymbol{u}$ and B is related to A by (1.8). Then the column of vectors defined by $\boldsymbol{v} = P\boldsymbol{u}$ is again a basis, and relative to this basis f has the matrix B, as we have seen. These results can be summed up as

Theorem 1.10

Let V be a finite-dimensional vector space over a field F, say $\dim V = n$. Then any linear transformation f of V is represented relative to a basis of V by an $n \times n$ matrix over F and the different matrices representing f in the different bases are all similar. ■

In general a matrix need not be square; it may have m rows and n columns, briefly $m \times n$, where m need not equal n. When $m \neq n$, the $m \times n$ matrices do not form a ring, but they still form an additive group; moreover an $m \times n$ matrix can be multiplied by an $r \times s$ matrix iff $n = r$ and the result is an $m \times s$ matrix.

Rectangular matrices can be used to describe homomorphisms between different vector spaces. Thus if U, V are vector spaces over a field F, of dimensions m, n respectively, then any linear mapping f from U to V is specified, relative to bases in these spaces, by an $m \times n$ matrix $A = (\alpha_{rs})$. Under a change of bases in the two spaces the matrix of f takes the form PAQ^{-1}, where P, Q are the matrices of transformation from the old to the new basis in U, V respectively. In every account of vector space theory it is shown that for a suitable choice of the invertible matrices P, Q we have $PAQ^{-1} = D = (d_{ij})$, where $d_{ij} = 1$ if $i = j = 1, \ldots, r$ for some r, and $d_{ij} = 0$ otherwise (see Exercise 1.2.11 below). This number r is known as the **rank** of A. The rank can also be defined as the maximum number of linearly independent rows of A, or equivalently, the maximum number of linearly independent columns of A. That these numbers are equal follows from the above definition, which was symmetric. For a symmetric definition of rank, applicable to matrices over any ring, let us write A, an $m \times n$ matrix, as a product:

$$A = PQ. \tag{1.9}$$

Here P is $m \times r$ and Q is $r \times n$, and the least value of r for the different factorizations of A is the rank of A. This follows because Eq. (1.9) may be interpreted as expressing the rows of A as a linear combination of the r rows of Q, or also as expressing the columns of A as a linear combination of the r columns of P. It follows that (i) for a given product expression (1.9) the rank of A cannot exceed r, and (ii) if the actual rank is s, then every row of A can be expressed as a linear combination of s rows, and this leads to a factorization (1.9) where Q has s rows; similarly we can use the columns of A. The rank so defined for matrices over general rings is also called the **inner rank**.

The $n \times n$ matrices of rank n are of particular importance, and we recall a result giving a number of equivalent conditions for this to happen.

Theorem 1.11

For any $n \times n$ matrix A over a field F the following conditions are equivalent:

(a) A has rank n;

(b) A has a right inverse, i.e. a matrix A' such that $AA' = I$;

(c) A has a left inverse, i.e. a matrix A'' such that $A''A = I$;

(d) for any matrix X with n rows, $AX = 0$ implies $X = 0$;

(e) for any matrix Y with n columns, $YA = 0$ implies $Y = 0$. ■

For the proof we refer to A.1, Chapter 5. We remark that when the left and right inverse both exist, they must be equal, for we have $A' = IA' = A''AA' = A''I = A''$. Of course these conditions are no longer equivalent for a rectangular matrix. Generally, any matrix A satisfying (d) is called **right regular**; if A does not satisfy (d) and is not 0, it is called a **left zero-divisor**. **Right zero-divisors** and **left regular** matrices are defined similarly, using (e); further, **regular** means "left **and** right regular" (in agreement with the earlier definition) and a **zero-divisor** is an element or matrix which is a left *or* right zero-divisor. Thus we see that by Theorem 1.11 a square matrix over a field is invertible iff it is (left or right) regular. We remark that these definitions apply to matrices over any ring, where they are used in particular for 1×1 matrices, i.e. elements.

The space of linear mappings from U to V will be denoted by $\mathscr{L}(U, V)$ or $\mathscr{L}_F(U, V)$ to indicate the field F. For the moment we shall only consider the special case where one of the spaces is the ground field F itself. A mapping α in $\mathscr{L}(F, V)$ is completely specified by the image of $1 \in F$, for if $1\alpha = a \in V$, then $x\alpha = x(1\alpha) = xa$; thus $\mathscr{L}(F, V) \cong V$, and this is an isomorphism if we regard both sides as vector spaces over F. In the same way we see that $\mathscr{L}(F^n, V) \cong V^n$, where V^n is the space of all rows of length n of vectors in V. We get a more interesting construction by taking the second argument equal to F. The space $\mathscr{L}(V, F)$ consists of all linear functions from V to F, also called **linear forms** on V. This space is denoted by V^* and is called the **dual** of V; if $\alpha \in V^*$, its value on $x \in V$ will be denoted by $\langle x, \alpha \rangle$. By definition this is a linear function of x, and it is also linear in α by the definition of $V^* = \mathscr{L}(V, F)$ as a linear space. Thus we have the rules

D.1 $\langle \lambda x + \mu y, \alpha \rangle = \lambda \langle x, \alpha \rangle + \mu \langle y, \alpha \rangle \qquad x, y \in V,\ \alpha \in V^*,\ \lambda, \mu \in F; \qquad (1.10)$

D.2 $\langle x, \lambda \alpha + \mu \beta \rangle = \lambda \langle x, \alpha \rangle + \mu \langle x, \beta \rangle \qquad x \in V,\ \alpha, \beta \in V^*,\ \lambda, \mu \in F. \qquad (1.11)$

We now show that the dual of V has the same dimension as V (at least in the finite-dimensional case, the only one considered here) and at the same time determine a basis for V^*.

Theorem 1.12

Let V be an n-dimensional space, with basis $v_1, \ldots, v_n$. Then its dual V^* is again n-dimensional, with a basis $\alpha_1, \ldots, \alpha_n$ such that $\langle v_i, \alpha_j \rangle = \delta_{ij}$, where δ_{ij} is the Kronecker delta.

Proof

Let $\alpha_1, \ldots, \alpha_n$ be the linear forms on Vdefined by the equations $\langle v_i, \alpha_j \rangle = \delta_{ij}$. The α_j are linearly independent, for if $\sum \xi_i \alpha_i = 0$, then for $i = 1, \ldots, n$, $\xi_i = \sum \xi_j \langle v_i, \alpha_j \rangle = \langle v_i, \sum \xi_j \alpha_j \rangle = 0$, so any linear relation between the α_j must be trivial, as claimed. To show that the α_j span V^* (and so form a basis), we need to write an arbitrary element λ of V^* as a linear combination of the α_i. Suppose that λ can be written as a linear combination of the α's:

$$\lambda = \sum \xi_i \alpha_i; \tag{1.12}$$

then on applying both sides to v_j we obtain $\langle v_j, \lambda \rangle = \sum \xi_i \langle v_j, \alpha_i \rangle = \xi_j$. So we have found values for ξ_j and it is easily checked that with these values (1.12) is satisfied. ■

The basis of V^* constructed above is called the **dual basis** and $\{v_1, \ldots, v_n\}$, $\{\alpha_1, \ldots, \alpha_n\}$ are **dual bases**. This result shows in particular that the dual of any finite-dimensional vector space is isomorphic to the space itself. However, the isomorphism depends on the choice of bases in the spaces and is not in any way natural, a point that will be elucidated in Section 1.5.

Exercises 1.2

1. Show that in any field F, the set $F \setminus \{-1\}$ is a group under the operation $(x, y) \mapsto x + y + xy$, with unit element 0.
2. Construct a field of four elements. (Hint: begin by constructing the multiplicative group.)
3. Show that in any ring all the regular elements have the same additive order; what are the values this order can take?
4. In an integral domain, if $ab = 1$, show that b is the inverse of a. Find an example of a general ring in which this is not true.
5. In any ring, if $aba = 1$, show that a has the inverse $ab (= ba)$.

6. Let M be an abelian group (with operation $+$) and suppose that there is an action of a field F on M satisfying V.1–V.3 but not V.4. Show that M can be written as a direct sum $M = M_0 \oplus M_1$, where M_1 also satisfies V.4, and so is an F-space, while the group M_0 has trivial F-action: $\alpha x = 0$ for all $\alpha \in F, x \in V$.

7. Let V be the space of all continuous real-valued functions on the unit interval. Show that V is a vector space over $\mathbb{R}$; show also that the differentiable functions form a subspace V_0 of V.

8. Let V, V_0 be as in Exercise 7 and for any $f \in V$ define $f^-(x) = f(-x)$, $f^1(x) = f(x+1)$, and for $f \in V_0$, let f' be the derivative. Which of the following mappings are linear? (i) $f \mapsto f^-$, (ii) $f \mapsto 1 + f^-$, (iii) $f \mapsto 2f^1$; on V_0, (iv) $f \mapsto f'$, (v) $f \mapsto (f')^2$, (vi) $f \mapsto f' + f^-$, (vii) $f \mapsto f - f^1$.

9. Show that the set of all polynomials in one indeterminate of degree at most n, with rational coefficients, form a vector space over $\mathbb{Q}$. What is its dimension?

10. Show that the set of all polynomials with rational coefficients form an infinite-dimensional vector space over $\mathbb{Q}$.

11. Let U, V be vector spaces over F, with bases given by the columns $\boldsymbol{u}$, $\boldsymbol{v}$ respectively. If under a change of basis $\boldsymbol{u}' = P\boldsymbol{u}, \boldsymbol{v}' = Q\boldsymbol{v}$, verify that a linear mapping from V to U whose matrix relative to $\boldsymbol{u}$, $\boldsymbol{v}$ is A has the matrix PAQ^{-1} relative to $\boldsymbol{u}'$, $\boldsymbol{v}'$. Deduce that in suitably adapted bases, a given mapping has the matrix $\operatorname{diag}(1, \ldots, 1, 0, \ldots, 0)$, where $\operatorname{diag}(a_1, \ldots, a_n)$ denotes the matrix with $a_1, \ldots, a_n$ along the main diagonal and all other elements equal to 0. Verify that the number of 1's on the main diagonal is just the rank of A.

12. Let A be an $m \times n$ matrix over a field. Show that the following conditions are equivalent and each implies $m \leq n$: (i) there exists A' such that $AA' = I$; (ii) $XA = 0$ implies $X = 0$; (iii) A has rank m.

1.3 Matrices

In Section 1.2 we met the ring $\mathfrak{M}_n(F)$ or F_n of all $n \times n$ matrices over a field F. Clearly this matrix ring can be defined over any ring R. Thus an $n \times n$ matrix over R is a family of n^2 elements of R, arranged as a square $P = (p_{ij})$, with the sum and product defined as in (1.7). The $n \times n$ matrices over R again form a ring R_n, but this is usually very different from the ring R. Thus R_n, for a non-zero ring and for $n > 1$, can never be an integral domain, because we have, for example,

$$\begin{pmatrix} 1 & 0 \\ 0 & 0 \end{pmatrix}\begin{pmatrix} 0 & 1 \\ 0 & 0 \end{pmatrix} = \begin{pmatrix} 0 & 1 \\ 0 & 0 \end{pmatrix}, \quad \begin{pmatrix} 0 & 1 \\ 0 & 0 \end{pmatrix}\begin{pmatrix} 1 & 0 \\ 0 & 0 \end{pmatrix} = 0.$$

These equations also show that R_n is not commutative for $n > 1$, even when R is commutative. Of course $0_n \cong 0$ and $R_1 \cong R$. As a further example we have the **scalar matrix** aI, in which all the diagonal entries are equal to a, while all other entries are 0; they form a subring of R_n isomorphic to R; thus R has been embedded in R_n. More generally, we have the **diagonal matrix** $\mathrm{diag}(a_1, \dots, a_n)$, which has the (i, i)-entry equal to a_i for $i = 1, \dots, n$ and all other entries zero.

More generally, we can consider matrices with m rows and n columns, i.e. $m \times n$ matrices, defined as families of mn elements of R, indexed by rows and columns as before, and with the same definitions of sum and product, but now we have to be careful to be sure that the matrices are conformable; this means that we can add two matrices that are both $m \times n$, but we cannot multiply them unless $m = n$. More generally, an $m \times n$ matrix can be multiplied by an $r \times s$ matrix precisely when $n = r$ and the result is then an $m \times s$ matrix. The set of all $m \times n$ matrices over a ring R will be denoted by ${}^mR^n$; as we saw, it is an additive group, but not a ring, unless $m = n$. If $m = 1$ or $n = 1$, the superscript 1 will be omitted; thus R^n is the space of n-component row vectors over R and ${}^mR$ the space of m-component column vectors. This notation is easily remembered by noting that we have product mappings ${}^mR^n \times {}^nR^p \to {}^mR^p$; in particular, ${}^mR^n \times {}^nR \to {}^mR$ and $R^m \times {}^mR^n \to R^n$.

We note a further operation on matrices, which has no analogue in rings. With each $m \times n$ matrix $P = (p_{ij})$ we associate an $n \times m$ matrix $P^{\mathrm{T}} = (p'_{ij})$, where $p'_{ij} = p_{ji}$. This is the matrix obtained by reflecting P in the main diagonal. It is called the **transpose** of P and satisfies the following rules when R is a commutative ring:

$$(P + Q)^{\mathrm{T}} = P^{\mathrm{T}} + Q^{\mathrm{T}}, (PQ)^{\mathrm{T}} = Q^{\mathrm{T}}P^{\mathrm{T}}, P^{\mathrm{TT}} = P. \tag{1.13}$$

It is clear that this operation of transposition is an anti-automorphism (in fact, an involution, i.e. an anti-automorphism of order two) of $\mathfrak{M}_n(R)$. The verification of these laws is straightforward and so is left to the reader. Here we note a point of notation which can lead to mistakes. Beginners sometimes think that the transpose of (p_{ij}) is the matrix (p_{ji}); in fact these two matrices are the same, because of the convention that the first suffix refers to rows and the second to columns, so the above form of the definition is needed.

Let R be any ring and for a given positive integer n, let R_n be the corresponding matrix ring. Then R can be embedded in R_n by the mapping $a \mapsto a.I$, where I is the unit matrix. There are other mappings, such as $a \mapsto ae_{11}$, where e_{11} is the matrix with 1 as (1,1)-entry and all other entries zero, but this is not a homomorphism (even though it is compatible with addition and multiplication) because it does not preserve the unit element. Let us more generally define e_{ij} as the matrix with (i, j)-entry 1 and all others 0. These elements are called the **matrix units** (rather perversely, because "units" are usually understood to be invertible elements); they satisfy the equations

$$e_{ij}e_{rs} = \delta_{jr}e_{is}, \qquad \sum e_{ii} = 1, \tag{1.14}$$

where δ is the Kronecker delta introduced in Section 1.2.

We have seen that the elements of R may be identified with the scalar matrices aI and these matrices commute with all the e_{ij}. What is of interest is that the converse holds: any ring containing n^2 elements e_{ij} satisfying Eqs. (1.14) is a matrix ring:

Theorem 1.13

Let R be a ring containing n^2 elements e_{ij} satisfying Eqs. (1.14). Then

$$R \cong \mathfrak{M}_n(C), \tag{1.15}$$

where the ring C is the centralizer of the e_{ij} in R.

Proof

Define C as in the statement and for $a \in R$ put $a_{ij} = \sum_r e_{ri}ae_{jr}$; then $a_{ij} \in C$, as is easily seen, and we have $\sum_{ij} a_{ij}e_{ij} = a$. Hence the bijection (1.15) is given by the correspondence $a \leftrightarrow (a_{ij})$, and this is easily verified to be an isomorphism. ∎

It is clear that the image of a set of matrix units under any ring homomorphism is again a set of matrix units. Hence Theorem 1.13 has the following useful consequence.

Corollary 1.14

Let $R = \mathfrak{M}_n(C)$ be a full matrix ring and $f : R \to S$ a homomorphism. Then S is again a full $n \times n$ matrix ring over the image of C under f. ∎

The concept of a matrix first occurs in the work of Arthur Cayley (1821–1895), although the idea of determinants goes back much further (Gottfried Wilhelm Leibniz, 1646–1716).

Exercises 1.3

1. Show that the ring of diagonal matrices in R_n is the direct sum of n ideals, each isomorphic to R.

2. A matrix A is called **full** if it is square, say $n \times n$, and its rank is n. Show that an $n \times n$ matrix with a $p \times q$ block of zeros cannot be full unless $p + q \leq n$.

3. Let A, B be two matrices with n columns, acting on the vector space ${}^nF$. Show that if $A\boldsymbol{u} = B\boldsymbol{u}$ for all $\boldsymbol{u} \in {}^nF$, then $A = B$; this follows even if $A\boldsymbol{u} = B\boldsymbol{u}$ holds for all vectors $\boldsymbol{u}$ in a basis of ${}^nF$.

4. Fill in the details of the proof of Theorem 1.13.

5. A matrix $A = (a_{ij})$ over a ring is called **upper (lower) triangular** if $a_{ij} = 0$ for $i > j$ $(i < j)$. Show that a triangular matrix is a unit iff all the diagonal elements are units.

6. Let $C = (c_{ij})$ be an $n \times n$ matrix such that $c_{ij} = 0$ for $i \geq j$ (such a matrix is **upper** 0-**triangular**). Verify that $C^n = 0$, but in general $C^{n-1} \neq 0$.

7. Let C be an $n \times n$ matrix over $\mathbb{R}$ such that $C^r = 0$ for some $r > n$. Show that there is an invertible matrix P such that $P^{-1}CP$ is upper 0-triangular; deduce that $C^n = 0$.

8. For any square matrix A define the **trace** as $\mathrm{tr}\ A = \sum a_{ii}$, where $A = (a_{ij})$. Show that $\mathrm{tr}\ AB = \mathrm{tr}\ BA$ (for matrices over a commutative ring). Show also that if $\mathrm{tr}\ AX = 0$ for all matrices X, then $A = 0$.

9. Let R be a ring with two elements u, v such that $u^2 = v^2 = 0$, $uv + vu = 1$. Show that R is a full 2×2 matrix ring over the centralizer of the set $\{u, v\}$.

10. Define $a = e_{11}, b = e_{12} + e_{23} + \ldots + e_{n1}$; show that $b^n = I$ and express all the matrix units e_{ij} as products $b^r a b^s$.

11. Let R be a ring with two elements a, b such that $b^n = 1 = \sum_{i=1}^{n} b^{i-1}ab^{1-i}$ and $ab^{i-1}a = \delta_{i1}a$ $(i = 1, \ldots, n)$. Verify that the elements $b^{i-1}ab^{1-j}$ form a set of matrix units.

12. Let R be a non-commutative ring and φ: $R \to R^\circ$ the anti-isomorphism. Verify that the mapping $A \mapsto (A\varphi)^{\mathrm{T}}$ is an anti-isomorphism from $\mathfrak{M}_n(R)$ to $\mathfrak{M}_n(R^\circ)$.

1.4 Modules

Vector spaces form an essential ingredient of linear algebra, which has been useful in formulating mathematical ideas in general terms. It is natural to look at the generalization in which fields are replaced by arbitrary rings. This leads to the notion of a module. Although modules do not share all the good properties of vector spaces, they give insight into the structure of rings and have many applications, particularly to the representation theory of groups.

Let A be an abelian group, written additively. As we saw in Section 1.1, the set $\mathrm{End}(A)$ of all endomorphisms of A forms a ring in a natural way. Since we often want to compare ring actions on different groups, it is convenient to define the action of a given ring on a group A. Given a ring R, we define an R-**module** as an abelian group M with a homomorphism from R to $\mathrm{End}(M)$. If we denote the effect of $r \in R$ on $x \in M$ by xr, the action satisfies the following rules, which recall the definition of a vector space in Section 1.2.

M.1 $(x + y)r = xr + yr$ for $r \in R$, $x, y \in M$;

M.2 $x(r + s) = xr + xs$ for $r, s \in R$, $x \in M$;

M.3 $x(rs) = (xr)s$ for $r, s \in R$, $x \in M$;

M.4 $x.1 = x$ for all $x \in M$.

Here we have written the action on the right of the terms acted on to ensure that the effect of rs is "first r, then s", as in M.3. In the commutative case (as in Section 1.2) it does not matter on which side the ring operators are written, but for general rings the side acted on is material. For this reason an R-module as defined above is called a **right** R-module. To emphasize the right module structure a right R-module M is written as M_R. By a **submodule** of M one understands any subgroup of the additive group which is closed under the operations of R.

Sometimes one wants to treat an abelian group M with a homomorphism from the opposite ring R° to $\mathrm{End}(M)$; this means that M.3 is to be replaced by $x(rs) = (xs)r$. Now it is more natural to write the operators on the left, and write M as ${}_RM$, so that M.3 takes the form

M.3° $(rs)x = r(sx)$ for $r, s \in R$, $x \in M$.

Of course the resulting module is called a **left** R-module. Summing up, we could restrict our attention to right modules (or to left modules) by allowing both R- and R°-modules, but it turns out to be more convenient to keep to R and use both left and right R-modules. When the side is immaterial, we shall speak of an "R-module"; when it is not necessary to specify the ring, we speak simply of a "module".

Given a module M and a subset X of M, we can again define the submodule **generated by** X as the set of all linear combinations $\sum x_i r_i$, where $x_i \in X$, $r_i \in R$; clearly it is the least submodule of M containing X. In particular, X is a generating set of M if every element of M is a linear combination of elements of X; if X can be chosen to be finite, M is said to be **finitely generated**. A family of elements $x_1, \ldots, x_n$ of a module is said to be **linearly independent** if $\sum x_i r_i = 0$ implies $r_i = 0$ for all i and **linearly dependent** otherwise. If $\langle x_i \rangle$ denotes the submodule generated by x_i, then the x_i are linearly independent precisely when the x_i are all non-zero and the sum of the $\langle x_i \rangle$ is direct. When M is a vector space over a field, linear dependence reduces to its usual meaning but whereas in a vector space in any linearly dependent family there is one vector dependent on the others, this need not be true in modules, e.g. in $\mathbb{Z}^2$, regarded as a $\mathbb{Z}$-module (cf. Exercise 1.4.2).

Sometimes we shall meet modules that are both left and right modules, such that the operations on the left and right commute. More precisely, given rings R, S, we define an R-S-**bimodule** as an abelian group M which is a left R-module and a right S-module such that

$$(rm)s = r(ms) \quad \text{for any } m \in M, r \in R, s \in S. \tag{1.16}$$

Here R and S may be the same ring, in which case we speak of an R-**bimodule**. For example, any ring R is an R-bimodule under the usual multiplication; in this case (1.16) is just the associative law in R.

As an obvious example of a module we take the ring $\mathbb{Z}$ of integers and any abelian group A, which becomes a $\mathbb{Z}$-module if we define for any $x \in A$, $r \in \mathbb{Z}$,

$$xr = \begin{cases} x + x + \ldots + x \ (r\,\text{terms}) & \text{if } r \geq 0, \\ -x(-r) & \text{if } r \leq 0. \end{cases}$$

We see that any subgroup of $\mathbb{Z}$ is a submodule.

As another example let us take a vector space V over some field k, with an endomorphism α of V. If $k[x]$ is the polynomial ring in x over k, then V may be regarded as a $k[x]$-module by defining for $v \in V$ and any polynomial $f(x) = \sum c_i x^i$, $vf(x) = \sum vc_i\alpha^i$. Here the submodules of V are the subspaces admitting α.

Given a ring R and two R-modules M and N, we define an R-**linear mapping** or **homomorphism** $f : M \to N$ as a homomorphism of the additive groups such that $(xf)r = (xr)f$ for all $x \in M$, $r \in R$. This just corresponds to a linear transformation of vector spaces. The set of all homomorphisms from M to N is denoted by $\mathrm{Hom}_R(M, N)$ or just $\mathrm{Hom}(M, N)$, when there is no risk of confusion. If $f, g \in \mathrm{Hom}(M, N)$, then we can define their sum $f + g$ by the rule $x(f + g) = xf + xg$, just as we form the sum of linear transformations of a vector space; in this way $\mathrm{Hom}(M, N)$ becomes an abelian group, as is easily checked. When R is commutative, $\mathrm{Hom}(M, N)$ also has an R-module structure, given by $x(fr) = (xr)f = (xf)r$, but this does not apply for general rings R; later, in Section 4.4, we shall see in what situations $\mathrm{Hom}(M, N)$ can be given a module structure.

It is easy to see that for any homomorphism $f : M \to N$ the image $\mathrm{im}\, f = \{y \in N \mid y = xf \text{ for some } x \in M\}$ is a submodule of N and the kernel, $\ker f = \{x \in M \mid xf = 0\}$ is a submodule of M. We observe that f is injective iff $\ker f = 0$. Moreover, the quotient group $M/\ker f$ can again be defined as an R-module and we have the **natural homomorphism** $\mathrm{nat} : M \to M/\ker f$ and the isomorphism $M/\ker f \cong \mathrm{im}\, f$, just as for abstract groups. The quotient $N/\mathrm{im}\, f$ is called the **cokernel** of f, written $\mathrm{coker}\, f$. Clearly f is surjective iff $\mathrm{coker}\, f = 0$. Generally we have for any submodule M' of M the **quotient module** M/M' obtained by putting the natural module structure on the quotient group M/M'. If we have submodules $A \supseteq B$ of M, we can form A/B, a quotient of the submodule A, also called a **factor module** of M. A homomorphism $f : M \to N$ may be illustrated by the following picture:

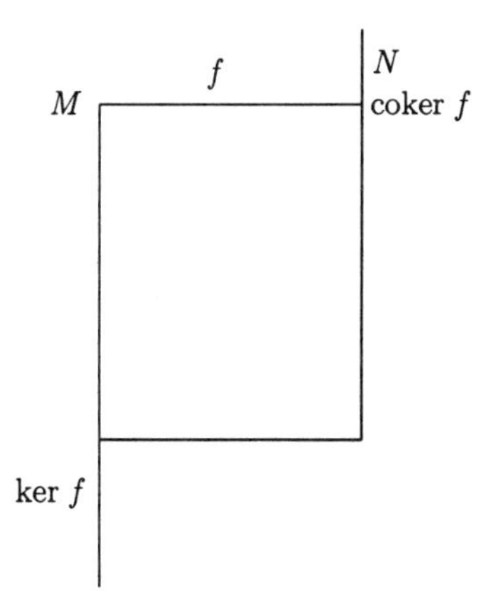

Later in this section we shall find a more useful representation, the exact sequence.

The above results may be stated as follows:

Theorem 1.15

Let R be a ring and M, N any R-modules. Given a homomorphism f: $M \to N$, its image im f is a submodule of N and its kernel ker f is a submodule of M. Conversely, given any submodule M' of M, an R-module structure may be defined on the set of cosets M/M' in such a way that the mapping nat : $M \to M/M'$ given by $a \mapsto a + M'$ is a surjective homomorphism with kernel M'.

Only the last part still needs proof. From group theory we know that the cosets M/M' form a group; to define a module structure, take a coset $x + M'$ and for $r \in R$ define $(x + M')r = xr + M'$. If x is replaced by $x + m$ where $m \in M'$, then $x + m + M' = x + M'$ and $mr \in M'$, hence $(x + m)r + M' = xr + mr + M' = xr + M'$, therefore the action of R on the cosets is well defined; now the module laws are easily verified and so we have an R-module M/M'; further, the mapping $a \mapsto a + M'$ is clearly a surjective homomorphism with kernel M'. ∎

As a useful consequence of Theorem 1.15 we have the **factor theorem**:

Corollary 1.16

Let R be a ring and let M, N be R-modules. Given a module homomorphism $f : M \to N$ and a submodule M' of M with natural homomorphism $\nu : M \to M/M'$, such that $M' \subseteq \ker f$, there is a unique mapping $f' : M/M' \to N$ such that f can be expressed as $f = \nu f'$. Moreover, f' is a homomorphism which is injective if and only if $M' = \ker f$.

The result is often expressed by saying that f can be **factored** by f'.

Proof

If such a mapping f'exists, it must satisfy

$$(x + M')f' = xf, \tag{1.17}$$

and this shows that there can be at most one such mapping. Since $M' \subseteq \ker f, xf$ is independent of its choice in the coset, so $xf = x'f$ whenever $x + M' = x' + M'$ and this means that the mapping f' given by (1.17) is well defined. Now it is easily verified that it is compatible with addition and the action of R. Thus f' is a homomorphism; the cosets mapped to 0 by f are just the cosets of M' in $\ker f$, so f' is injective precisely when $M' = \ker f$. ■

For groups we have the isomorphism theorems, which are basic in any treatment. They have analogues for modules (and indeed any algebraic systems), which are proved in the same way. Below we state the results with brief indications of the proofs.

Theorem 1.17 (First Isomorphism Theorem)

Let R be any ring. Given R-modules M, N and a homomorphism $f : M \to N$, let $\nu : M \to M/\ker f$ be the natural homomorphism and $\lambda : \operatorname{im} f \to N$ the inclusion mapping. Then f has a factorization $f = \nu f'\lambda$, where $f' : M/\ker f \to \operatorname{im} f$ is an isomorphism.

Proof

By the factor theorem (Corollary 1.16) we have $f = \nu g$ for some homomorphism $g : M/\ker f \to N$, and since the image of g is clearly $\operatorname{im} f$, we can write $g = f'\lambda$ for some f'; here f' is a homomorphism from $M/\ker f$ to $\operatorname{im} f$, thus it is one-one and onto and hence an isomorphism. ■

Of particular importance is the **Second Isomorphism Theorem**, also called the **parallelogram rule**:

Theorem 1.18

Let M be a module and A, B two submodules of M. Then $A + B = \{x + y | x \in A, y \in B\}$ and $A \cap B$ are again submodules of M and we have an isomorphism

$$A/(A \cap B) \cong (A + B)/B.$$

The proof is as for groups: we have to check that the mapping $x \mapsto x + B$ of A into M/B is a homomorphism with kernel $A \cap B$ and image $(A + B)/B$; the details may be left to the reader. ■

To describe quotients of quotients we have the **Third Isomorphism Theorem**:

Theorem 1.19

Let M be a module and M' a submodule. Then the submodules of M/M' correspond in a natural way to the submodules of M containing M' and if N/M' is the submodule corresponding to $N \supseteq M'$, then $(M/M')/(N/M') \cong M/N$.

Proof

The natural homomorphism from M/M' to M/N is surjective, with kernel N/M'. It follows by Theorem 1.17 that $M/N \cong M/M'/N/M'$. ∎

The isomorphism theorems also have an analogue for rings, which we state below. First we note the factor theorem:

Theorem 1.20

Given rings R, S and a homomorphism $f : R \to S$, and any ideal A of R such that $A \subseteq \ker f$, there exists a unique map $f' : R/A \to S$ such that the triangle below is commutative. Moreover, f' is a homomorphism which is injective if and only if $A = \ker f$:

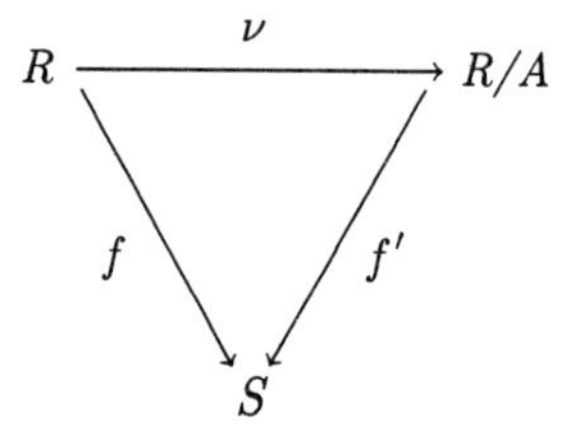

∎

We now have the analogues of Theorems 1.17–1.19, with the same proofs, which will therefore not be repeated.

Theorem 1.21 (First Isomorphism Theorem)

Given rings R, S and a homomorphism $f : R \to S$, let $\nu : R \to R/\ker f$ be the natural homomorphism and $\lambda : \operatorname{im} f \to S$ the inclusion mapping. Then f has a factorization $f = \nu f' \lambda$, where $f' : R/\ker f \to \operatorname{im} f$ is an isomorphism. ∎

Theorem 1.22 (Second Isomorphism Theorem)

Let R be a ring, R' a subring and A an ideal of R. Then $A \cap R'$ is an ideal of R' and there is an isomorphism

$$R'/(A \cap R') \cong (R' + A)/A. \qquad \blacksquare$$

Theorem 1.23 (Third Isomorphism Theorem)

Let R be a ring and A an ideal of R. Then the subrings (resp. ideals) of R/A correspond in a natural way to subrings (resp. ideals) of R that contain A and if B/A is the ideal of R/A corresponding to the ideal $B \supseteq A$ of R, then $(R/A)/(B/A) \cong R/B$. ■

To give an example, let us first verify that a field F has no ideals apart from F itself and 0. If A is an ideal $\neq 0$, then A contains a non-zero element c; so it also contains $1 = c.c^{-1}$. Hence it contains every element of $F: x = x.1 \in A$, whence $A = F$, as we had to show. In general a ring is called **simple** if it is non-zero and has no ideals apart from itself and 0. It may easily be verified that every field is simple. An ideal A of a ring R is said to be **maximal** if $A \neq R$ and there are no ideals lying properly between A and R. Now Theorem 1.23 shows that in a commutative ring R an ideal A is maximal iff R/A is a field. Of course a non-commutative ring may well be simple without being a field; we shall meet examples later.

As an important example of an R-module we have the ring itself. It is clear from M.1–M.4 that any ring R is a right R-module under the ring multiplication. A submodule of R is a subgroup A of the additive group such that $AR \subseteq A$; this is called a **right ideal**. Similarly, regarding R as a left R-module, we find that the submodules are the **left ideals**, defined as subgroups B of the additive group of R such that $RB \subseteq B$. A right ideal which is also a left ideal is just an ideal in the sense defined earlier (in Section 1.1), also called a **two-sided ideal**. Of course in a commutative ring the notions of left, right and two-sided ideal coincide.

Frequently we have a sequence of modules with a homomorphism from each module to the next:

$$\dots \xrightarrow{r_0} M_0 \xrightarrow{r_1} M_1 \xrightarrow{r_2} M_2 \xrightarrow{r_3} \dots . \tag{1.18}$$

This sequence is said to be **exact** at M_1 if $\text{im } r_1 = \ker r_2$. The sequence is **exact** if it is exact at every module. Thus the simplest exact sequences are of the form $0 \to M \to 0$, possible only if $M = 0$, and $0 \to M \to M' \to 0$, which indicates that $M \cong M'$. These are rather trivial cases; the first non-trivial case is a sequence of the form

$$0 \to M' \xrightarrow{\alpha} M \xrightarrow{\beta} M'' \to 0. \tag{1.19}$$

An exact sequence of this form is called a **short exact sequence**. It means that M has a homomorphism onto M'' with kernel isomorphic to M' and (1.19) is often used in the analysis of a homomorphism.

As an example of (1.19), suppose that M' is a submodule of M with a complement N isomorphic to M''; this means that every element a of M can be written in just one way as $a = b + c$, where $b \in M', c \in N$. In this case M is said to be the **direct sum** of M' and N and we write $M = M' \oplus N$. We now have an exact sequence of the form (1.19), where α is the inclusion mapping and β is the isomorphism from N to M'', but this is a rather special case of an exact sequence, where the image of M' is a direct summand in M; such a sequence is said to be **split exact**. It means that there are homomorphisms $\lambda : M'' \to M$ such that $\lambda\beta$ is the identity on M'' and $\mu : M \to M'$ such that $\alpha\mu$ is the identity on M'; conversely, either of these conditions will ensure that (1.19) is split exact.

If we compare modules with vector spaces we see that the two concepts have much in common; in fact, vector spaces are just the special case of modules over a field, but there are also important differences. Thus every vector space has a basis (as we saw in Section 1.2 in the finite-dimensional case); this greatly simplifies their theory and there is no direct counterpart in the general case. However, for any ring there are also modules which possess a basis; they are the free modules, defined as follows.

Let R be any ring; an R-module F is said to be **free** on a set X if there is a mapping $i : X \to F$ such that, given any mapping $f : X \to M$ to an R-module M, there exists a unique homomorphism $f' : F \to M$ such that $f = if'$. The definition becomes a little clearer if we use a diagram; F is free on X iff there is a unique f' making the diagram shown below commutative.

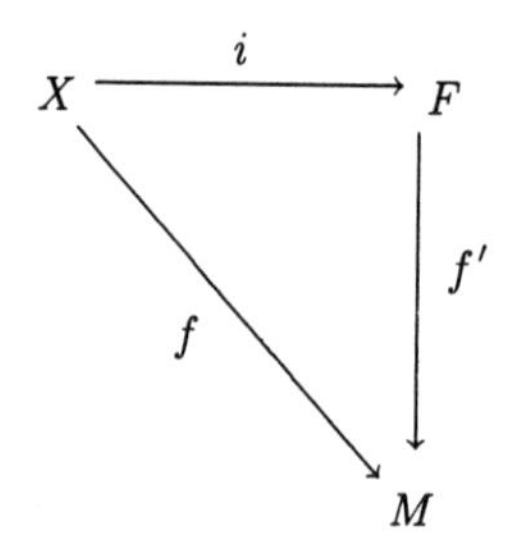

The property singled out in the above definition is called the **universal mapping property**, and F is called a **universal object** for mappings $f : X \to M$. It is the most important property of free modules, but it does not give us a way of constructing free modules, nor even assuring us that they exist. To carry out a construction, and hence an existence proof, we shall limit ourselves at first to the case of main concern to us, when X is finite, say $X = \{x_1, \ldots, x_n\}$. Let us form the set F of all linear combinations $\sum x_i r_i = x_1 r_1 + \ldots + x_n r_n$, where $r_i \in R$ and $\sum x_i r_i = 0$ implies $r_i = 0$ for all i. This set F is an R-module with the operations $\sum x_i r_i + \sum x_i s_i = \sum x_i (r_i + s_i)$, $\sum (x_i r_i) s = \sum x_i (r_i s)$; to verify the universal mapping property, take any R-module M and a mapping $f : X \to M$. If a homomorphism $f' : F \to M$ with the required property is to exist, then, since for any element u of F there is just one expression $u = \sum x_i r_i$, we must have $uf' = (\sum x_i r_i) f' = \sum (x_i r_i) f' = \sum (x_i f) r_i$ and this determines the image of u uniquely. Thus F is the free R-module on X. In the general case, when X is not necessarily finite, we may write it as $X = (x_i), i \in I$, where I is an index set. The construction is now the same as before, with the convention that all sums are finite, i.e. any element of F is a linear combination of finitely many of the x_i; this is usually expressed by writing the element as a sum $\sum x_i r_i$ as before with the understanding that all but a finite number of the r_i are zero; more briefly, the r_i are **almost all** zero. We sum up the result in:

Theorem 1.24

Let R be any ring. Then for any set X there exists a free R-module on X. This module is determined up to isomorphism by X and has the universal mapping property. ■

This result presents an aspect of free modules that shows a resemblance to vector spaces. Nevertheless, free modules do not share all the good properties of the latter; thus each vector space has a well-defined dimension, which indicates the number of elements in a basis. A free R-module also has a basis, by definition, but there is nothing to show that all bases of a given free module have the same number of elements. However, this is true at least for commutative rings. Here we understand by a **basis** of a free R-module a linearly independent generating set.

Theorem 1.25

Let F be a finitely generated free module over a non-trivial commutative ring R. Then all bases of F are finite and have the same number of elements.

Proof

By hypothesis F is free, on a basis $\{f_i | i \in I\}$. Further, it is finitely generated, by $u_1, \ldots, u_n$ say. Each of $u_1, \ldots, u_n$ can be expressed in terms of finitely many of the f_i, so F is generated by finitely many of the f_i, hence there can only be finitely many and this shows the basis to be finite.

Now let $\boldsymbol{e} = (e_1, \ldots, e_m)$, $\boldsymbol{f} = (f_1, \ldots, f_n)$ be two rows representing different bases of F, with transformations $\boldsymbol{f} = \boldsymbol{e}P$, $\boldsymbol{e} = \boldsymbol{f}Q$ between them. It follows that $\boldsymbol{e}PQ = \boldsymbol{f}Q = \boldsymbol{e}$, i.e. $PQ = I$, and similarly, $QP = I$, where P is $m \times n$ and Q is $n \times m$. If $m \neq n$, suppose that $m < n$ and extend P, Q to form square matrices by adding $n - m$ rows of zeros to P and $n - m$ columns of zeros to Q:

$$P_1 = \begin{pmatrix} P \\ 0 \end{pmatrix}, \qquad Q_1 = (Q \quad 0);$$

since $Q_1 P_1 = QP = I$, it follows that $\det Q_1 . \det P_1 = 1$, but $\det Q_1 = 0$, because Q_1 has a zero column, and this is in contradiction with the fact that $1 \neq 0$ in R. ■

Remarks

1. We note that the trivial ring 0 has to be excluded, since the theorem is not true in that case: with a literal interpretation one finds that over the trivial ring every module is zero.
2. Since determinants are defined only for commutative rings, the proof does not apply to general rings. In fact, the result is not true for all rings, though it holds for most of the rings encountered below (see Section 4.7).
3. The result still holds for free modules that are not finitely generated, but we shall not need this fact (see A.2, p. 142).

The number of elements in a basis of a free module F is called the **rank** of F.

Given any finitely generated R-module M, let X be a generating set and denote the free module on X by F. Then the identity mapping on X can be extended to a homomorphism $f : F \to M$, which is surjective, because X was chosen as a generating set. Thus we have an exact sequence

$$0 \to K \xrightarrow{i} F \xrightarrow{f} M \to 0, \tag{1.20}$$

where K is the module of relations. This is called a **presentation** of M; if K is finitely generated as well as F, M is said to be **finitely presented**. In case only K is finitely generated, M is said to be **finitely related**. In general the relation module K need not be free and by presenting it as the image of a free module and continuing the process, we obtain a **free resolution** of M, whose study forms part of homological algebra (see A.3, Chapter 3 and Section 4.6 below).

We have already met the direct sum construction; we recall that a module M is the direct sum of the submodules N_1, N_2, written

$$M = N_1 \oplus N_2, \tag{1.21}$$

if every element of M is expressible as a sum of an element of N_1 and one of N_2 in just one way. Equivalently, (1.21) holds precisely when $M = N_1 + N_2$ and $N_1 \cap N_2 = 0$. In this case N_1 and N_2 are said to be **complemented** and each is a **complement** of the other.

Given any two modules M, N (over the same ring R), we can form a module which is the direct sum of M and N. We define a module L as the set of all pairs (x, y), $x \in M, y \in N$, with componentwise operations:

$$(x,y) + (x',y') = (x + x', y + y'), \qquad (x,y)\alpha = (x\alpha, y\alpha)$$

$$x, x' \in M,\ y, y' \in N,\ \alpha \in R. \tag{1.22}$$

This direct sum L is again denoted by $M \oplus N$, although the meaning is slightly different from (1.21). To be precise, what we have formed here is known as the **external** direct sum, in contrast to the **internal** direct sum (1.21). The mappings $\lambda : x \mapsto (x, 0)$, $\mu : y \mapsto (0, y)$ are clearly injective, and if the images are denoted by M', N' respectively, then we can express L as an internal direct sum: $L = M' \oplus N'$.

It is clear how the direct sum of a finite family of modules is to be defined; given R-modules $M_1, \ldots, M_r$, their direct sum $M = M_1 \oplus \ldots \oplus M_r$ is the set of all r-tuples with the i-th component in M_i and elementwise operations, as in (1.22). This construction can be carried out even for an infinite family $(M_\lambda)(\lambda \in \Lambda)$, but now there are two possibilities: We can form a module P whose elements are all families (x_λ) with $x_\lambda \in M_\lambda$, or we can take the subset S consisting of the families almost all of whose components are zero. It is easily checked that S is again an R-module; it is called the **direct sum** of the M_λ, written $\oplus_\lambda M_\lambda$, or also $\coprod_\lambda M_\lambda$, while P is their **direct product**, denoted by $\prod_\lambda M_\lambda$. The direct product contains, for each λ, a submodule isomorphic to M_λ, obtained as the set of elements whose components are all zero, except possibly for the λ-component. These submodules, taken together, generate the direct sum S, which is thus embedded in P.

These modules P and S have a universal property which is often useful. For each λ there is a projection mapping $\pi_\lambda : P \to M_\lambda$ with the property that for any family of homomorphisms $f_\lambda : A \to M_\lambda$ from an R-module A there is a unique homomorphism $f : A \to P$ such that $f_\lambda = f\pi_\lambda$ for all $\lambda \in \Lambda$; we simply define for any $a \in A$, $af = (af_\lambda)$. This property is expressed by the isomorphism

$$\mathrm{Hom}_R\left(A, \prod_\lambda M_\lambda\right) \cong \prod_\lambda \mathrm{Hom}_R(A, M_\lambda). \tag{1.23}$$

Similarly for the direct sum we have injections $\mu_\lambda : M_\lambda \to S$ such that for any family of homomorphisms $g_\lambda : M_\lambda \to B$ to an R-module B there exists a unique homomorphism $g : S \to B$ such that $g_\lambda = \mu_\lambda g$ for all $\lambda \in \Lambda$. This yields the corresponding isomorphism

$$\mathrm{Hom}_R\left(\coprod_\lambda M_\lambda, B\right) \cong \prod_\lambda \mathrm{Hom}_R(M_\lambda, B). \tag{1.24}$$

Exercises 1.4

1. Show that every abelian group can be defined as a $\mathbb{Z}$-module in a unique way. Under what conditions can an abelian group be defined as a $\mathbb{Z}/(m)$-module?

2. State the condition for two non-zero vectors in $\mathbb{Z}^2$ to be linearly dependent and give an example of two vectors that are linearly dependent, though neither is linearly dependent on the other.

3. Given a ring homomorphism $f : R \to S$, show that every S-module can be considered as an R-module in a natural way.

4. Given two R-modules M, N and two homomorphisms f, g from M to N, show that the subset $\{x \in M | xf = xg\}$ of M is a submodule.

5. Let M, N be R-modules and X a generating set of M. Given $f, g \in \mathrm{Hom}_R(M, N)$, if $xf = xg$ for all $x \in X$, show that $f = g$. (Hint: use Exercise 4.)

6. Given a module homomorphism $f : M \to N$, define ker f, im f as before, and further define the **coimage** as coim $f = M/\ker f$. Derive the following commutative diagram with exact row:

$$\begin{array}{ccccccccc} 0 \to \ker f \to & M & \xrightarrow{f} & N & \to \operatorname{coker} f \to 0 \\ & \downarrow & & \uparrow & \\ & \operatorname{coim} f & \longrightarrow & \operatorname{im} f & \end{array}$$

Show that the bottom row is an isomorphism.

7. Show that every finitely related module can be written as a direct sum of a finitely presented module and a free module.

8. Show that every short exact sequence of the form (1.19) in which M'' is free, is split exact. Deduce that every short exact sequence of vector spaces is necessarily split exact.

9. Verify the equivalence of the two definitions of (external) direct sum given above.

10. Let A be a cyclic group of order 2 and B a cyclic group of order 3. Verify that their direct sum $A \oplus B$ is cyclic of order 6. What is the general rule?

1.5 The Language of Categories

Category theory is a large and growing subject and it would be quite unreasonable (as well as unnecessary) to expect readers of this book to be familiar with it, but the language of categories can be very useful in gaining an understanding of ring theory. This section is therefore devoted to a brief account of this topic.

In Section 1.1 we met a class of objects, called "rings", with certain mappings between them, called "homomorphisms". Most readers will have met the class of "groups" with their homomorphisms, and it turns out to be fruitful to put this idea in abstract terms. A **category** $\mathscr{A}$ is a collection of objects; for each pair of $\mathscr{A}$-objects A, B there is a set denoted by $\mathrm{Hom}(A, B)$ or $\mathrm{Hom}_{\mathscr{A}}(A, B)$ or even $\mathscr{A}(A, B)$, whose members are called the **$\mathscr{A}$-morphisms** (or just **morphisms**) from A to B. Instead of $f \in \mathrm{Hom}(A, B)$ we also write $f : A \to B$. We say "f goes from A to B" and call A the **source** of $f$and B its **target**; the source and target are uniquely determined by the morphism f. With each pair of morphisms $f : A \to B$, $g : B \to C$ there is associated a morphism $h : A \to C$, called the **composite** or **product** of f and g and denoted by fg; thus fg goes from the source of f to the target of g, and it is defined iff the target of f coincides with the source of g. We note that in some books (where mappings are habitually written on the left), the composite of f and g is denoted by gf; we shall not follow this practice, except in special cases, where this is dictated by circumstances.

The morphisms satisfy the following laws:

C.1 If $f : A \to B$, $g : B \to C$, $h : C \to D$, so that $(fg)h$ and $f(gh)$ are both defined, then they are equal:

$$(fg)h = f(gh). \tag{1.25}$$

C.2 With each object A a morphism $1_A : A \to A$, called the **identity morphism** on A, is associated such that for any $f : A \to B$, $g : C \to A$,

$$1_A f = f, g 1_A = g. \tag{1.26}$$

Thus 1_A acts as a left neutral for morphisms with source A and as right neutral for morphisms with target A. In particular, $\text{Hom}(A, A)$ is just a semigroup with unit element 1_A, i.e. a monoid. There can only be one element 1_A satisfying (1.26), for if $1'$ is a second one, then $1' = 1'1_A = 1_A$. Given $f : A \to B$, $g : B \to A$, if $fg = 1_A, gf = 1_B$, then g is called the **inverse** of f; it is easily verified that if an inverse of f exists, it is uniquely determined by f. It is denoted by f^{-1} and f is called an **isomorphism** if it has an inverse; f is called an automorphism if it is an isomorphism for which source and target coincide.

Examples

1. The collection of all sets forms a category Ens (for the French for "set": ensemble) with all mappings as morphisms, with the usual composition rule; thus fg means: first f, then g. We remark that the actual image of a mapping f may well be smaller than the target of f, as morphism. This example makes it clear why category theory speaks of a "collection" or "class" of objects, rather than a set: we cannot very well speak of the "set of all sets" without running into difficulties, which are discussed in books on set theory. For this reason Ens is sometimes called a **large** category; in a **small** category the object class forms a set.
2. The class of all groups, with all homomorphisms as morphisms, forms a category Gp. Strictly speaking, we should also specify the composition rule for morphisms, but as in the previous example this is just "composition of mappings". We note that an isomorphism between groups may also be defined as a homomorphism which is bijective.
3. The class of all topological spaces, with continuous mappings as morphisms. Here it is no longer true that every bijective morphism is an isomorphism, since it may change the topology. For example, for any topological space T, if T^* denotes the same set with the discrete topology, then the identity mapping from T^* to T is a bijective morphism which is not an isomorphism, unless T is also discrete.

4. Let S be a partially ordered set (see the remarks on notation). This set becomes a category if we define $\mathrm{Hom}(x,y)$ to have a single element $\bullet$ if $x \leq y$ and to be empty otherwise. Composition is defined by $\bullet\bullet = \bullet$; with these rules we have a category which is small (some people have called this a kittygory).
5. For a given ring R form a category Mat whose objects are the positive integers; the morphisms from m to n are $m \times n$ matrices over R, with matrix multiplication as composition.
6. For any ring R we can form the category with a single object R, whose members are the morphisms, with the ring multiplication as composition. Like Examples 4 and 5, this is a small category.

We see that usually, though not always, the morphisms in a category are mappings, with the usual composition rule. We shall use diagrams to illustrate morphisms, as for mappings. For example, given $f : A \to B$, $g : B \to C$, $h : A \to C$, if $h = fg$, the diagram below is said to **commute**, or to be **commutative**.

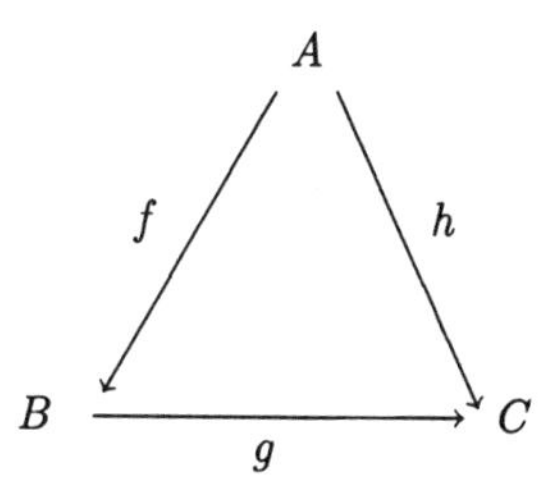

Given a category $\mathscr{A}$, by a **subcategory** of $\mathscr{A}$ we understand a subclass $\mathscr{A}'$ of the object class of $\mathscr{A}$, which is a category whose set of morphisms from A to B is a subset of $\mathrm{Hom}_{\mathscr{A}}(A,B)$, with the same composition as in $\mathscr{A}$. Thus we have

$$\mathrm{Hom}_{\mathscr{A}'}(A,B) \subseteq \mathrm{Hom}_{\mathscr{A}}(A,B), \tag{1.27}$$

and of course the morphism sets of $\mathscr{A}'$ are closed under composition. If in (1.27) we have equality for all objects in $\mathscr{A}'$, then $\mathscr{A}'$ is said to be a **full** subcategory of $\mathscr{A}$. Thus for a full subcategory we need only specify the objects, not the morphisms. For example, fields form a full subcategory of the category of rings; on the other hand, vector spaces over a given field F form a subcategory of the category of abelian groups, which is not full.

Given categories $\mathscr{A}$, $\mathscr{B}$, if $\mathscr{A}$ is a subcategory of $\mathscr{B}$, the inclusion mapping may be regarded as a structure-preserving morphism between categories, known as a functor. Generally we define a **functor** from $\mathscr{A}$ to $\mathscr{B}$ as a function T which assigns to each $\mathscr{A}$-object X a $\mathscr{B}$-object X^T or XT and to each $\mathscr{A}$-morphism $f : X \to Y$ a $\mathscr{B}$-morphism $f^T \colon X^T \to Y^T$, subject to the laws:

F.1 If fg is defined in $\mathscr{A}$, then $(f^T)(g^T)$ is defined in $\mathscr{B}$ and

$$f^T.g^T = (fg)^T.$$

F.2 For each $\mathscr{A}$-object X we have $1_X^T = 1_{X^T}$

Sometimes we have a functor which reverses the multiplication in F.1; again with each $\mathscr{A}$-object X a $\mathscr{B}$-object X^T is associated, and with each $\mathscr{A}$-morphism f from X to Y there is now associated a $\mathscr{B}$-morphism f^T from Y^T to X^T, satisfying F.2, while F.1 is replaced by

F.1* If fg is defined in $\mathscr{A}$, then $g^T.f^T$ is defined in $\mathscr{B}$ and

$$g^T.f^T = (fg)^T.$$

A functor is called **covariant** if it satisfies F.1, **contravariant** if it satisfies F.1*.

Examples of functors

1. If we omit the unit element from the definition of rings, we obtain a generalization, satisfying only R.1–R.6 and not R.7, which in Section 1.1 we called a "rung". It is clear that rungs form a category, with structure-preserving mappings as morphisms, while rings form a subcategory. This subcategory is not full, because a morphism between rings, qua rungs, need not preserve the unit element. This inclusion of categories is a functor, which in effect forgets the special role played by the 1 of the ring. Many functors have this form of a "forgetful" functor, as in the next example.
2. Let Rg be the category of rings and homomorphisms, and with each ring associate its underlying additive group. This is a functor from Rg to the category Ab of all abelian groups, which consists in "forgetting" the multiplicative structure of the ring.
3. If R, S are rings, then any ring homomorphism from R to S may be regarded as a functor from R to S, both regarded as categories with a single object.
4. Given any object C in a category $\mathscr{A}$, there is a functor h^C from $\mathscr{A}$ to Ens, which associates with each $\mathscr{A}$-object X the set $\text{Hom}(C, X)$ and with each morphism $f : X \to Y$ associates the mapping $\text{Hom}(C, X) \to \text{Hom}(C, Y)$ which makes $\alpha : C \to X$ correspond to $\alpha f : C \to Y$. It is clear that for $Y = X$ and $f = 1_X$ we obtain the identity mapping $\alpha \mapsto \alpha 1_X$. The other functor rule states that $fh^C.gh^C = (fg)h^C$, i.e. $(\alpha f)g = \alpha(fg)$, and this follows from the commutative diagram

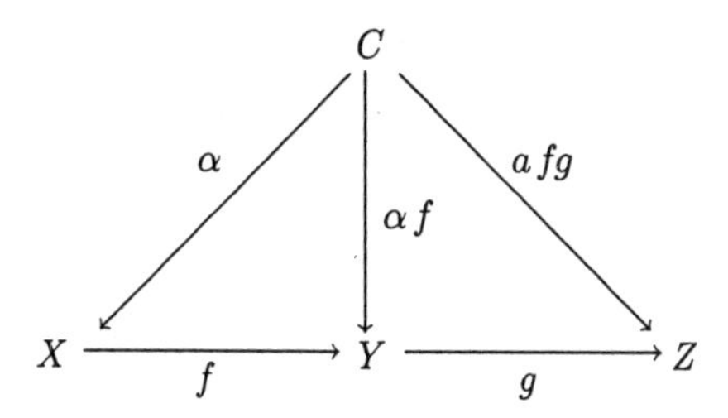

This functor h^C is called the **covariant hom functor**; it is an important example that will recur frequently. From (1.23) we see that the covariant hom functor preserves direct products, at least in the case of modules, though with appropriate definitions this can be shown to hold quite generally.

5. If $\mathscr{A}$ and C are as in Example 4, then there is a contravariant functor h_C, called the **contravariant hom functor**, associating with each $\mathscr{A}$-object X the hom-set $\mathrm{Hom}(X, C)$ and with each morphism $f : X \to Y$ the mapping $\mathrm{Hom}(Y, C) \to \mathrm{Hom}(X, C)$ which makes $\gamma : Y \to C$ correspond to $f\gamma : X \to C$. Again we have $1_X h_C = h_C$, and by reversing all the arrows in the diagram, we see that $(gf)\gamma = g(f\gamma)$, i.e. $gf.h_C = fh_C.gh_C$, where we have to remember that $g(f\gamma)$ means: first apply f, then apply g, to γ. By (1.24), the contravariant hom functor converts direct sums to direct products.
6. Let Rg be the category of rings (and homomorphisms) and Mat_n the category of pairs consisting of a matrix ring with a chosen set of matrix units. We have a functor α from Rg to Mat_n which associates with any ring R its $n \times n$ matrix ring R_n and its set of matrix units, and a functor β from Mat_n to Rg which associates with a given pair (S, e_{ij}) the centralizer subring of the matrix units in S. At this point a natural question arises: Is it true to say that $\alpha\beta = 1$ or $\beta\alpha = 1$, where 1 is the identity functor on the appropriate category? The answer is: Almost but not quite: we have isomorphism, not identity; the details will be given below.
7. Let V be a vector space over a field F; we have seen that the space of linear forms, $V^* = \mathscr{L}(V, F)$, is again a vector space, the dual of V, and the correspondence from V to V^* is a functor. Any linear mapping $f : U \to V$ induces the mapping $f^* : V^* \to U^*$ which associates with $\alpha \in V^*$ the element $f^*\alpha$ of U^*; this means that the effect of $f^*\alpha$ on $u \in U$ is the same as that of α on uf, or in symbols,

$$\langle uf, \alpha \rangle = \langle u, f^* \alpha \rangle.$$

It is not hard to verify that the correspondence from f to f^* is a contravariant functor. Let $\boldsymbol{v} = (v_1, \ldots, v_n)$, $\alpha = (\alpha_1, \ldots, \alpha_n)$ be a pair of dual bases for V and V^*; thus $\langle v_i, \alpha_j \rangle = \delta_{ij}$. If $\boldsymbol{v}'$, α' is a second pair of dual bases, say $v_i' = \sum p_{ir} v_r$, $\alpha_j' = \sum q_{js} \alpha_s$, then $\delta_{ij} = \langle v_i', \alpha_j' \rangle = \sum p_{ir} q_{js} \delta_{rs} = \sum p_{ir} q_{jr}$, or on writing $P = (p_{ir})$, $Q = (q_{js})$, we have $PQ^T = I$. Thus if the basis of V is transformed by P, that of V^* is transformed by the transpose of the inverse of P. Strictly speaking, we are here dealing, not with the category of vector spaces, but with the category $\mathscr{V}$, say, of pairs $(V,\boldsymbol{v})$ consisting of a vector space V and a basis $\boldsymbol{v}$ of V; the functor * we have defined is contravariant, from $\mathscr{V}$ to itself.

In any category $\mathscr{A}$ an **initial object** is an object I such that for any $\mathscr{A}$-object X, $\mathrm{Hom}(I, X)$ consists of a single element. For example, in the category of all sets, Ens, the empty set $\varnothing$ is an initial object, for, given any set X, there is exactly one mapping of $\varnothing$ into X; this may sound strange at first, but if we remember that a mapping is essentially a set of instructions where to send the elements of the source, we see that there is only one way of doing this, which consists in doing nothing (because there are no elements in $\varnothing$). As another example in the category of groups, Gp, the one-element group $\{e\}$ is an initial object, because in any homomorphism e has to map to the unit element of the target group. We note that an initial object is unique up to isomorphism, because if I, I' are two initial objects and f, g the unique morphisms from I to $I'$and I' to I, then $fg \in \mathrm{Hom}(I, I)$, and this hom-set has only one member, which must be the identity on I. Similarly gf is the identity on I'. Dually we define a **final object** as an object F such that from any object there is just one morphism to F. More briefly we could define a final object of $\mathscr{A}$ as an initial object in the opposite category $\mathscr{A}^\circ$. We see that the trivial group is also a final object in Gp. An object which is both initial and final is called a **zero object**; thus the trivial group is a zero object in Gp. We remark that every category can be enlarged by a zero object in just one way (see Exercise 1.5.9).

Let $\mathscr{A}$, $\mathscr{B}$ be any two categories and S, T covariant functors from $\mathscr{A}$ to $\mathscr{B}$. A **natural transformation** from S to T is defined as a family of $\mathscr{B}$-morphisms $\varphi_X : XS \to XT$ for each $\mathscr{A}$-object X such that for any $\mathscr{A}$-morphism $f : X \to Y$ the diagram shown below commutes:

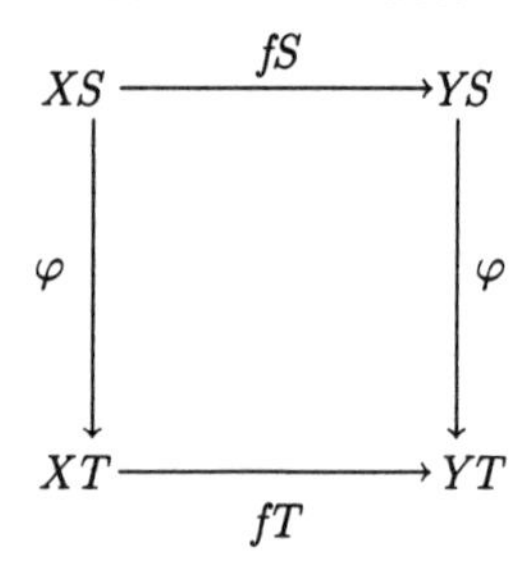

Thus $fS\varphi_Y = \varphi_X fT$. If the transformation φ has a two-sided inverse, which is again a natural transformation, φ is called a **natural isomorphism**; a corresponding definition applies for contravariant functors S, T. Natural transformations from a covariant to a contravariant functor are much rarer and will not be needed here.

Let $\mathscr{A}$, $\mathscr{B}$ be any categories and let S be a functor from $\mathscr{A}$ to $\mathscr{B}$; this functor is called an **isomorphism** of categories if it has an inverse, i.e. a functor T from $\mathscr{B}$ to $\mathscr{A}$ such that ST is the identity on $\mathscr{A}$ and TS is the identity on $\mathscr{B}$. Such functors are quite rare; for instance in Example 6 above, if we take a ring R, form the $n \times n$ matrix ring R_n with matrix units e_{ij} and then take the centralizer of these matrix units, we obtain a ring isomorphic to R, but not R itself. It may seem pedantic to insist on this kind of distinction, but it makes things run more smoothly; in this sense category theory is rather like working a computer, where attention to detail is equally necessary.

Guided by the above example, we define two categories $\mathscr{A}$, $\mathscr{B}$ to be **equivalent** if there is a functor $S : \mathscr{A} \to \mathscr{B}$ and a functor $T : \mathscr{B} \to \mathscr{A}$ such that ST is naturally isomorphic to the identity functor on $\mathscr{A}$ and similarly TS is naturally isomorphic to the identity functor on $\mathscr{B}$. When S and T are contravariant, $\mathscr{A}$ and $\mathscr{B}$ are called **anti-equivalent** or **dual**. For example, any category $\mathscr{A}$ is dual to its **opposite**, that is the category $\mathscr{A}^\circ$ which has the same objects and morphisms as $\mathscr{A}$, but for each morphism source and target are interchanged and composition is in the opposite order. Thus to a morphism $f : X \to Y$ in $\mathscr{A}$ there corresponds a morphism $f^\circ : Y \to X$ in $\mathscr{A}^\circ$. We see that the matrix functor discussed in Example 6 above is an equivalence between Rg and Mat_n and the correspondence from V, f to V^*, f^* for finite-dimensional vector spaces is a duality.

It is useful to have a criterion for recognizing a natural isomorphism of functors. To find one we observe that any covariant functor $S : \mathscr{A} \to \mathscr{B}$ defines for each pair of $\mathscr{A}$-objects X, Y a mapping

$$\mathrm{Hom}_{\mathscr{A}}(X, Y) \to \mathrm{Hom}_{\mathscr{B}}(X^S, Y^S). \tag{1.28}$$

If this mapping is surjective, S is said to be **full**; if it is injective, S is called **faithful**, and S is called **dense**, if every $\mathscr{B}$-object is isomorphic to one of the form X^S, for some $\mathscr{A}$-object X. Now we have

Proposition 1.26

A functor between two categories is an equivalence if and only if it is full, faithful and dense.

Proof

For an equivalence $S : \mathscr{A} \to \mathscr{B}$, (1.28) is a bijection, so S is then full and faithful; it is also dense, for if ST and TS are naturally isomorphic to identities, then every $\mathscr{B}$-object Y is isomorphic to Y^{TS}. Conversely, assume that S is full, faithful and dense. Then (1.28) is a bijection, and hence an isomorphism. Further for each $\mathscr{B}$-object Y there exists an $\mathscr{A}$-object Y^T, say, such that $Y^{TS} \cong Y$, and we can use the isomorphism (1.28) to transfer any morphism between $\mathscr{B}$-objects to one between the corresponding $\mathscr{A}$-objects, and this shows S to be an equivalence. ■

To give an example of a natural transformation which is not an equivalence, consider the category of groups, Gp. We define two functors on it, associating with each group G, firstly G^{ab}, the group made abelian; this is the commutator quotient G/G', where G' is the derived group, generated by all commutators $(x, y) = x^{-1}y^{-1}xy$, for $x, y \in G$. Secondly, we take the group modulo squares G/G^2, where G^2 is the subgroup generated by all squares x^2, $x \in G$. It is clear that we have here two functors from Gp to Ab, the category of all abelian groups. That G^{ab} is abelian is clear, because $(x, y) = 1$ by definition. The same holds for G/G^2, because $(x, y) = x^{-2}(xy^{-1})^2y^2 = 1$; this also shows that $G^2 \subseteq G'$, so we have a homomorphism from G^{ab} to G/G^2, and this is a natural transformation, in the sense that the square below commutes:

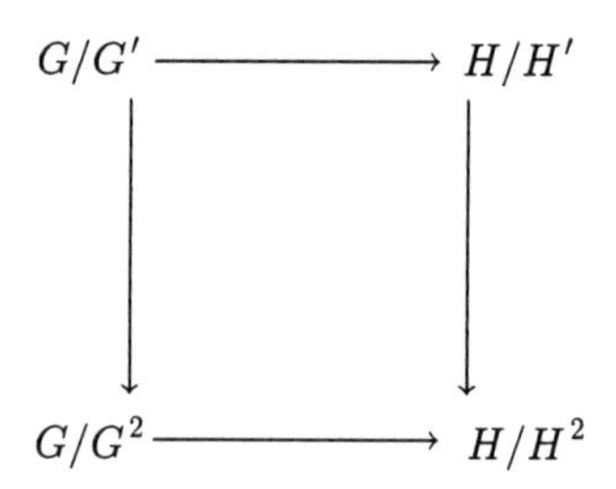

As a second example, in which there occurs a natural isomorphism, consider a finite-dimensional vector space V and its dual space V^*. We have seen in Section 1.2 that V and V^* are isomorphic, but this isomorphism depends on the choice of bases in the two spaces and is not natural. However, when we come to consider the bidual of V, V^{**}, we find that there is a mapping from V to V^{**}, which associates with any element x of V the linear form on V^* (= element of V^{**}) defined by

$$\alpha \mapsto \langle \alpha, x \rangle,$$

and it is easily verified that this is a natural isomorphism; here the finite dimensionality is essential. It follows that the functor * is a duality of the category of finite-dimensional vector spaces with itself. In fact it was this natural duality which led Eilenberg and Mac Lane in 1942 to introduce the notions of functor and category (Samuel Eilenberg, 1913–1998, Saunders Mac Lane, 1909–).

Exercises 1.5

1. For any set S define $\mathscr{P}(S)$ as the set of all subsets of S. Verify that the correspondence from S to $\mathscr{P}(S)$ which assigns to $f : S \to T$ the mapping $\mathscr{P}(S) \to \mathscr{P}(T)$ taking a subset X of S to its image set Xf in T is a covariant functor. Verify that the correspondence which assigns to $f : S \to T$ the mapping from $\mathscr{P}(T)$ to $\mathscr{P}(S)$ given by $X \mapsto f^{-1}X$ (the inverse image of X) is a contravariant functor.

2. Verify that in any category the inverse of a morphism f from A to B, when it exists, is uniquely determined and goes from B to A.

3. Show that $\mathbb{Z}$ is an initial object in the category of rings and 0 is a final object.

4. If a partially ordered set is interpreted as a category (as in the text), what are the initial and final objects (if they exist)?

5. Let $\mathscr{A}$ be any category and in each class of isomorphic objects choose one object. Verify that the chosen objects form a full subcategory of $\mathscr{A}$; this is called the **skeleton** of $\mathscr{A}$. Does Ens have a small skeleton (i.e. is the class of isomorphism classes of objects a set)?

6. Show that the category of all finite sets and mappings has a skeleton isomorphic to $\mathbb{N}$.

7. Let C be any small category and $\mathscr{A}$ an arbitrary category. Show that the functors from C to $\mathscr{A}$ form the object class of a category whose morphisms are the natural transformations.

8. In any category an object which is both initial and final is called a **zero object**. Verify that between any two objects there is just one morphism which goes via the zero object, the **zero morphism**. Which of the following categories has a zero object? Ens, Gp, Rg, Top.

9. Let $\mathscr{A}$ be any category and define $\mathscr{A}''$ as the category obtained from A by adjoining one object Z with a single morphism $Z \to A$ and a single morphism $A \to Z$ for each object A of $\mathscr{A}$, as well as a single morphism $Z \to Z$, and with the natural compositions. Verify that $\mathscr{A}''$ is a category with zero.

10. Let $\mathscr{A}$ be a category and P an object in $\mathscr{A}$. Define a new category $(P, \mathscr{A})$, the **comma category** determined by P, whose objects are morphisms $\mu_X : P \to X$ $(X \in \mathscr{A})$ and whose morphisms are morphisms $f : X \to Y$ such that $\mu_Y = \mu_X f$. Show that this category has an initial object.

2
Linear Algebras and Artinian Rings

The study of linear algebras is important for the theory of group representations, with many applications in physics, as well as being a topic of independent interest. It is usually subsumed under the topic of Artinian rings, since many of the proofs carry over to this class. This chapter brings the main results of the theory, the Wedderburn theorems, and explains the role of the radical, and as an application, includes a brief introduction to group representations.

2.1 Linear Algebras

Many rings have a field as operator domain and this leads to the following definition. Let k be a field; by a **linear algebra** over k, also called a k-**algebra**, we understand a ring A with a mapping from $k \times A$ to A, denoted by $(\alpha, r) \mapsto \alpha r$, such that

LA.1 $\alpha(r+s) = \alpha r + \alpha s$, for all $\alpha \in k$, $r, s \in A$;

LA.2 $(\alpha+\beta)r = \alpha r + \beta r$, for all $\alpha, \beta \in k$, $r \in A$;

LA.3 $(\alpha\beta)r = \alpha(\beta r)$, for all $\alpha, \beta \in k$, $r \in A$;

LA.4 $1.r = r$, for all $r \in A$;

LA.5 $\alpha(rs) = (\alpha r)s = r(\alpha s)$, for all $\alpha \in k$, $r, s \in A$.

We see that LA.1–LA.4 just tell us that a linear algebra is a (left) k-module, while LA.5 describes the effect of α on a product in A. As a k-module A is in fact a vector space, with a well-defined dimension, which is frequently assumed to be finite, and the reader may want to keep this case in mind, though we shall not always impose this restriction.
Sometimes one wishes to consider algebras without a unit element; generally we shall call A **non-unital** if the existence of a unit element is not assumed.

As a vector space over k, A has a basis $\{u_i\}$; this means that every element a of A can be written in the form of a finite sum $a = \sum a_i u_i$, where $a_i \in k$ and almost all the a_i (i.e. all but a finite number) are zero, in case the basis is infinite. Further, the coefficients a_i are uniquely determined by a. The multiplication in A is completely determined by the products of the basis elements. Thus we have the equations

$$u_i u_j = \sum_k \gamma_{ijk} u_k, \tag{2.1}$$

where the elements γ_{ijk} of k are called the **multiplication constants** of the algebra; in case A is infinite-dimensional, it is of course understood that only finitely many of the γ_{ijk} are non-zero for any choice of i, j. Once (2.1) is given, the product of $a = \sum a_i u_i$ and $b = \sum b_j u_j$ is defined by linearity:

$$ab = \sum a_i b_j \gamma_{ijk} u_k.$$

It is clear that these multiplication constants still depend on the choice of the basis $\{u_i\}$; if we change the basis to $\{v_r\}$, where $v_r = \sum a_{ri} u_i$, with a matrix $A = (a_{ri})$, whose inverse is $A^{-1} = (\alpha_{ir})$, then the multiplication in terms of the v_r reads

$$v_r v_s = \sum_t \eta_{rst} v_t, \quad \text{where} \quad \eta_{rst} = \sum_{ijk} a_{ri} a_{sj} \alpha_{kt} \gamma_{ijk}, \tag{2.2}$$

as is easily verified.

Since the multiplication in A is associative, the multiplication constants cannot be chosen arbitrarily but have to satisfy the equations arising from the equation $(u_i u_j) u_k = u_i (u_j u_k)$, that is,

$$\sum_l \gamma_{ijl} \gamma_{lkm} = \sum_l \gamma_{jkl} \gamma_{ilm}. \tag{2.3}$$

Similarly, when the algebra is commutative, we have

$$\gamma_{ijk} = \gamma_{jik}.$$

Occasionally one may also wish to consider non-associative algebras, where (2.3) is not satisfied, but we shall not have occasion to do so except in examples.

Let us give some examples of linear algebras, where k is any field:

1. Matrix algebra. For any $n \geq 1$, the ring of all $n \times n$ matrices over k forms a linear algebra, with basis given by the matrix units e_{ij}. The algebra is n^2-dimensional, with multiplication $e_{ij}e_{kl} = \delta_{jk}e_{il}$, where δ_{jk} is the Kronecker delta.
2. Group algebra. Let G be a group. Then we can form the group algebra kG of G over k, by taking G as a basis of our algebra. Thus the elements of kG are the finite sums $\sum \alpha_g g$, where $g \in G$, $\alpha_g \in k$, with multiplication $(\sum \alpha_g g)(\sum \beta_h h) = \sum \alpha_g \beta_h gh$. Put more simply, kG has G as k-basis and the product in kG is given by the multiplication table of G.
3. Monoid algebra. The same construction can be carried out with a monoid M instead of a group, leading to the monoid algebra kM. As a simple example we have the polynomial ring $k[x]$, formed from the infinite monoid on x (see Section 3.1).
4. Quaternions. The first genuine skew field was found by Hamilton in 1843 (Sir William Rowan Hamilton, 1805–1865). This ring may be described as the algebra $\mathbb{H}$ over $\mathbb{R}$ with basis 1, i, j, k and multiplication $\mathrm{ij} = -\mathrm{ji} = \mathrm{k}, \mathrm{jk} = -\mathrm{kj} = \mathrm{i}, \mathrm{ki} = -\mathrm{ik} = \mathrm{j}$. Thus the elements have the form $u = a + b\mathrm{i} + c\mathrm{j} + d\mathrm{k}$; if $\bar{u} = a - b\mathrm{i} - c\mathrm{j} - d\mathrm{k}$, then $\bar{u}u = a^2 + b^2 + c^2 + d^2$, and this is a positive real number unless $u = 0$. It follows that $\mathbb{H}$ is a skew field.
5. Trivial algebra. Let V be any vector space over k and define $A = k \oplus V$ as an algebra by the rule $(\alpha, u).(\beta, v) = (\alpha\beta, 0)$. We obtain an algebra in which the product of any two elements of V is zero. If we allow non-unital algebras ("rungs"), then V itself would be an algebra, albeit a rather trivial one.
6. More generally, the definition of a k-algebra allows k to be any commutative ring. The algebra A is said to be a **faithful** k-algebra if the mapping $k \longrightarrow A$ given by $\alpha \mapsto 1$ is injective; of course this condition is automatically satisfied d when k is a field.

Although we shall mainly be concerned with associative linear algebras in this book, we shall give some non-associative examples for illustration; as it happens, most of them are of importance in applications.

7. Lie algebra. A linear algebra L is called a **Lie algebra** if it is not assumed to be associative or have a unit element, but satisfies instead the identities

$$x^2 = 0, \tag{2.4}$$

$$(xy)z + (yz)x + (zx)y = 0. \tag{2.5}$$

Equation (2.5) is called the **Jacobi identity** (after Carl Gustav Jacob Jacobi, 1804–1851). By expanding the identity $(x + y)^2 = 0$ and using (2.4) we obtain the equation $xy = -yx$. Lie algebras arise as the "infinitesimal groups" associated with continuous groups ("Lie groups"); they are named after Sophus Lie (1842–1899).

8. Jordan algebra. A linear algebra A is called a **Jordan algebra** if it is not assumed to be associative but satisfies the identities

$$xy = yx, \tag{2.6}$$

$$(xy)y^2 = (xy^2)y. \tag{2.7}$$

They are named after Pascual Jordan (1902–1980) and are used in the study of quantum mechanics.

9. Algebras in genetics. By a **gametic algebra** one understands a linear algebra A of finite dimension n, say, whose multiplication constants satisfy the relations

$$0 \leq \gamma_{ijk} \leq 1,$$

$$\sum_{k=1}^{n} \gamma_{ijk} = 1 \qquad \text{for all } i, j = 1, \ldots, n, \tag{2.8}$$

$$\gamma_{ijk} = \gamma_{jik} \qquad i, j, k = 1, \ldots, n.$$

Such algebras are used in the study of genetics. Given a population of diploid individuals (i.e. having chromosomes in homologous pairs), suppose that $u_1, \ldots, u_n$ are the genetically distinct gametes (i.e. germ cells) produced; then the state of the population can be described by a vector $(\alpha_1, \ldots, \alpha_n)^{\mathrm{T}}$, where the gamete frequencies α_i satisfy

$$0 \leq \alpha_i \leq 1 \quad (i = 1, \ldots, n), \qquad \sum_{i=1}^{n} \alpha_i = 1 \tag{2.9}$$

Such a state can be characterized by a vector $\sum \alpha_i u_i$, and these vectors form a gametic algebra, where the multiplication constant γ_{ijk} is proportional to the number of gametes of type u_k produced by a union of the gametes u_i, u_j (i.e. the zygote $u_i u_j$), with the proportionality factor chosen so that (2.8) holds. Of course in the case of sex-linked inheritance the final equations in (2.8) need not hold (because the algebra may be non-commutative).

For a given k-algebra A consider a right A-module M. We can identify k with a subalgebra of A by the rule $\alpha \mapsto \alpha.1$ and in this way M becomes a k-space. Moreover, every homomorphism between A-modules is k-linear, as is easily verified.

Given a finite-dimensional algebra A over a field k, of dimension n, say, consider the right multiplication by an element $a \in A$:

$$\rho_a : x \mapsto xa, \qquad \text{where } x \in A. \tag{2.10}$$

This is a linear transformation of A and so can be represented by an $n \times n$ matrix over k. Thus we have a mapping $\rho : A \to k_n$, and this is easily seen to be a homomorphism. Such a homomorphism into a full matrix ring is called a **matrix representation** or simply a **representation** of A. The right multiplication ρ_a is called the **regular representation** of A. More precisely, it is the **right** regular representation; the **left regular representation** is defined similarly by left multiplication:

$$\lambda_a : x \mapsto ax, \qquad \text{where } x \in A. \tag{2.11}$$

Strictly speaking, this is not a representation, but an **anti**-representation, for it is easily seen that we have $\lambda_{ab} = \lambda_b\lambda_a$.

Let us describe ρ_a explicitly in terms of a basis $u_1, \ldots, u_n$ of A. If $a = \sum \alpha_i u_i$, then ρ_a maps u_i to $u_i a = \sum_j u_i \alpha_j u_j = \sum_{jk} \alpha_j \gamma_{ijk} u_k$, therefore ρ_a is represented by the matrix

$$(\rho_a)_{ij} = \sum_k \alpha_k \gamma_{ikj}, \tag{2.12}$$

relative to the basis $\{u_i\}$. This representation is **faithful**, i.e. with zero kernel, for if $\rho_a = 0$, then $a = 1.a = 1\rho_a = 0$. Thus we have proved

Theorem 2.1

Let A be an n-dimensional algebra over a field k. Then A is isomorphic to a subalgebra of the matrix ring $\mathfrak{M}_n(k)$. ■

Exercises 2.1

1. Let A be a k-algebra, where k is a field. Given an A-module M, verify that any endomorphism of M as A-module is a k-linear mapping.
2. Verify the formula (2.2) for the multiplication constants under a change of basis.
3. Show that the centre of a simple algebra is a field.
4. Verify that the regular representation (2.10) is a homomorphism of A as right A-module. How is the matrix $(\rho_{ij}(a))$ affected by a change of basis?

5. Let A be a k-algebra without a unit element and define A^1 as an algebra on the vector space $k \oplus A$ with the multiplication $(\alpha, a)(\beta, b) = (\alpha\beta, \alpha b + \beta a + ab)$. Show that A^1 is an algebra with unit element $(1, 0)$ and A as a (non-unital) subalgebra. If A has a unit element, what is its relation to the unit element in A^1?

6. Prove an analogue of Theorem 2.1 for non-unital algebras using Exercise 5.

7. Let A be any associative algebra and define an algebra A^- on the space A by taking as the product $[x, y] = xy - yx$. Show that A^- is a Lie algebra (it can be shown that every Lie algebra can be obtained from an associative algebra in this way).

8. Let A be any algebra and define a **derivation** of A as a linear mapping δ of A into itself such that $(xy)\delta = x\delta.y + x.y\delta$. Show that the set $\Delta(A)$ of all derivations on A is a linear space and in fact an algebra, though not in general associative, but that $\Delta(A)^-$ (defined as in Exercise 7) is always a Lie algebra.

9. A derivation δ (as in Exercise 8) is said to be **nilpotent** if $\delta^{n+1} = 0$ for some integer n. If δ is a nilpotent derivation on an algebra A, say $\delta^{n+1} = 0$, show that $\exp \delta = 1 + \delta + \delta^2/2! + \ldots + \delta^n/n!$ is an automorphism of A, and find its inverse.

10. Show that the three-dimensional vector space $V = \mathbb{R}^3$ forms an associative algebra with respect to the operation $x \uparrow y = (x_1y_1, x_2y_2, x_3y_3)$. Using the shift operator S defined by $xS = (x_2, x_3, x_1)$, find an expression in terms of $\uparrow$ for the vector product $(x \wedge y)\wedge z$ and deduce that V with the product $x \wedge y$ is a Lie algebra.

11. Let A be any associative algebra over a field of characteristic not 2, and define an algebra A^+ on the space A by taking as product $\{x, y\} = xy + yx$. Verify that A^+ is a Jordan algebra (it can be shown that not every Jordan algebra is of this form; those that are, are called **special**).

12. Let A, A^+ be as in Exercise 11. Show how to express the operation $(x, y) \mapsto xyx$ in terms of $\{x, y\}$.

2.2 Chain Conditions

In studying the structure of a ring, its ideals, right, left and two-sided, are of great importance. Of course any general results on left ideals follow from the corresponding results on right ideals by symmetry; more generally, both types of results are subsumed under the theory of modules, and it is most convenient to frame general definitions for modules. Given a module M, we shall frequently be concerned with a **chain** of submodules C_i, i.e. a sequence of the form

$$C_1 \subseteq C_2 \subseteq \ldots \subseteq C_r \qquad \text{or} \qquad C_1 \supseteq C_2 \supseteq \ldots \supseteq C_r, \tag{2.13}$$

known as an **ascending** chain and a **descending** chain respectively. The case where all such chains are finite is of particular interest. For example, this is always the case when M is a finite-dimensional vector space. More generally, a module is said to be **Artinian** (after Emil Artin, 1898–1962) if in every descending chain of submodules we have equality from some point onwards, or in other words, every descending chain of submodules becomes stationary; the module is called **Noetherian** (after Emmy Noether, 1882–1935) if every ascending chain of submodules becomes stationary. For example, $\mathbb{Z}$ considered as a module over itself, is Noetherian but not Artinian. Our first aim is to find other equivalent ways of expressing these definitions. We recall that in any non-zero module M a **maximal submodule** is a proper submodule N such that $N \subseteq M' \subseteq M$ implies that $M' = N$ or $M' = M$; a **minimal submodule** is a non-zero submodule N such that $M' \subseteq N$ implies $M' = 0$ or $M' = N$. Clearly a minimal submodule is **simple**, i.e. it is non-zero, with no non-zero proper submodules.

Theorem 2.2

For any module M the following conditions are equivalent:

(a) M is Noetherian;

(b) every strictly ascending chain of submodules in M breaks off after a finite number of steps;

(c) any non-empty collection of submodules of M has a maximal member;

(d) every submodule of M is finitely generated.

Proof

(a) $\Rightarrow$ (b). Since (a) holds, we have in any ascending chain equality after a finite number of terms, but in a strictly ascending chain equality is excluded, so the chain must break off. (b) $\Rightarrow$ (c). When (b) holds and $\mathscr{C}$ is a non-empty collection of submodules of M, define N_1 to be any member of $\mathscr{C}$. Either N_1 is maximal or $\mathscr{C}$ contains N_2 which strictly contains N_1. Continuing in this way, we obtain a strictly ascending chain $N_1 \subset N_2 \subset \dots$, which must break off, by (b), but this can only happen when we have found a maximal element. (c) $\Rightarrow$ (d). If N is a submodule of M, let $\mathscr{C}$ be the collection of all finitely generated submodules and choose a maximal term N' in $\mathscr{C}$. If $N' \subset N$, we can adjoin an element to N' to obtain N'' in $\mathscr{C}$ and properly containing N', but this contradicts the maximality of N'. Hence $N' = N$ and this shows N to be finitely generated. Finally, to prove that (d) $\Rightarrow$ (a), assume that we are given an ascending chain of submodules $N_1 \subseteq N_2 \subseteq \dots$, put $N = \bigcup N_i$ and let $a_1, \dots, a_r$ be a finite generating set of N. If $a_j \in N_{i_j}$ and $k = \max\{i_1, \dots, i_r\}$, then equality holds in our chain from N_k onwards. ■

Conditions (a)–(c) have obvious analogues for Artinian modules, proved in the same way, so we obtain

Corollary 2.3

For any module M the following conditions are equivalent:

(a) M is Artinian;

(b) every strictly decreasing chain breaks off after a finite number of steps;

(c) any non-empty collection of submodules of M has a minimal member. ■

For a finitely generated module we have the following characterization:

Proposition 2.4

A module M is finitely generated if and only if the union of every ascending chain of proper submodules is proper.

Proof

Suppose that M is finitely generated, by $u_1, \dots, u_n$, say. If we have an ascending chain of submodules $C_1 \subset C_2 \subset \dots$, whose union is M, then for $i = 1, \dots, n$, $u_i \in C_{r_i}$ and if $r = \max\{r_1, \dots, r_n\}$, then $C_r = \sum u_i R = M$, so not all the C_i are proper. Conversely, if M is not finitely generated, we obtain an infinite strictly ascending chain of proper submodules by setting $C_0 = 0$ and if C_n has been defined and is finitely generated, then $C_n \neq M$, because M is not finitely generated; hence we can find $a_{n+1} \in M \setminus C_n$ and writing $C_{n+1} = C_n + a_{n+1}R$, we obtain a finitely generated submodule $C_{n+1} \supset C_n$. By induction we thus have an infinite strictly ascending chain of finitely generated (hence proper) submodules, which can be continued indefinitely. ■

Strictly speaking what we need is a notion of transfinite induction to establish this result for general modules, but this is not needed in what follows. It would take us rather far afield to introduce this notion and would not really be useful, so the reader should bear in mind that (a) the general result can be proved rigorously, and (b) the above proof is adequate for any module with a countable generating set.

Given a chain of submodules of M, we may be able to insert terms to obtain a **refinement** of our chain, i.e. a chain with more links; when this is not possible, our original chain is said to be a **composition series**. Thus a series (2.13) is a composition series for M iff $C_r = M$ and each factor C_i/C_{i-1} is a simple module. In that case r, the number of simple factors, is called the **composition length** or simply the **length** of M, denoted by $\ell(M)$. As we shall soon see (in Theorem 2.9), if a module M has a composition series, then all its composition series have the same length; more generally, any two isomorphic modules which have composition series have the same length. Now for any homomorphism $f : M \to N$, we have the isomorphisms $M/\ker f \cong \operatorname{im} f$, $\operatorname{coker} f = N/\operatorname{im} f$. It follows that $\ell(M) - \ell(\ker f) = \ell(\operatorname{im} f)$, $\ell(\operatorname{coker} f) = \ell(N) - \ell(\operatorname{im} f)$, and so we obtain the equation

$$\ell(M) - \ell(\ker f) = \ell(N) - \ell(\operatorname{coker} f),$$

whenever all terms are defined.

The following result is an easy consequence of this equality.

Theorem 2.5

Let M, N be two modules of finite lengths and let $f : M \to N$ be a homomorphism between them. If f is injective, then $\ell(M) \leq \ell(N)$, while if f is surjective, then $\ell(M) \geq \ell(N)$, with equality (in both places) if and only if f is an isomorphism. ■

Two chains of submodules in a module M:

$$M = C_0 \supseteq C_1 \supseteq \ldots \supseteq C_r = 0$$

and

$$M = D_0 \supseteq D_1 \supseteq \ldots \supseteq D_s = 0$$

are said to be **isomorphic** if they have the same length: $r = s$ and the factors are isomorphic in pairs, thus for some permutation $i \mapsto i'$ of $(1, \dots, r)$ we have

$$C_{i-1}/C_i \cong D_{i'-1}/D_{i'}.$$

Our aim will be to prove that any two composition series of a module are isomorphic. To accomplish this aim, we shall need a relation between submodules, known as the **modular law**:

Theorem 2.6

Let M be any module. Given any submodules A, B, C, where $A \subseteq C$, we have the relation

$$A + (B \cap C) = (A + B) \cap C. \tag{2.14}$$

Further, if $A + B = C + B, A \cap B = C \cap B$, then $A = C$.

Proof

It is clear that $A \subseteq A + B$ and $B \cap C \subseteq C$, hence the left-hand side of (2.14) is contained in the right-hand side. Conversely, if z lies in the right-hand side, say $z = x + y$, where $x \in A, y \in B$, then it follows that $x \in C$ and so $y = z - x \in C$. Thus $y \in B \cap C$, and hence $z = x + y \in A + (B \cap C)$, as claimed in (2.14). When the further equations are satisfied, we can rewrite (2.14) as $A + (A \cap B) = (B + C) \cap C$, which reduces to $A = C$. ∎

The last point can be interpreted as saying that any two comparable submodules with the same complement are equal, in other words: the complements of a given submodule (if any) are pairwise incomparable.

Our next object is to compare different chains of submodules in a given module M. In order to do so we shall need the Zassenhaus lemma (Hans Zassenhaus, 1912–1991), which compares two factors in M and is based on the parallelogram rule (Theorem 1.18):

Lemma 2.7

Let M be any module with submodules $C' \subseteq C$, $D' \subseteq D$. Then we have

$$\frac{D' + (C \cap D)}{D' + (C' \cap D)} \cong \frac{C \cap D}{(C' \cap D) + (C \cap D')} \cong \frac{C' + (C \cap D)}{C' + (C \cap D')}. \tag{2.15}$$

Proof

By applying the parallelogram rule to $D'$and $C \cap D$, we obtain

$$[D' + (C \cap D)]/D' \cong (C \cap D)/(C \cap D'). \tag{2.16}$$

In this isomorphism the submodule $(C \cap D') + (C' \cap D)$ corresponds to $D' + (C \cap D') + (C' \cap D)$, which reduces to $D' + (C' \cap D)$; hence by the third isomorphism theorem we find

$$\frac{D' + (C \cap D)}{D' + (C' \cap D)} \cong \frac{C \cap D}{(C' \cap D) + (C \cap D')}.$$

This proves the first isomorphism of (2.15); the second follows by symmetry. ■

The factor $[D' + (C \cap D)]/[D' + (C' \cap D)]$ is called the **projection** of the factor C/C'on D/D'; thus the lemma tells us that the projection of any factor on a second factor is isomorphic to the projection of the second factor on the first. We now take any two chains of submodules in M and project their factors on each other:

$$M = C_0 \supseteq C_1 \supseteq \ldots \supseteq C_r = 0, \tag{2.17}$$

$$M = D_0 \supseteq D_1 \supseteq \ldots \supseteq D_s = 0. \tag{2.18}$$

If we define

$$C_{ij} = (D_j \cap C_{i-1}) + C_i, \qquad i = 1, \ldots, r, \quad j = 0, \ldots, s,$$

then we obtain a chain from C_{i-1} to C_i:

$$C_{i-1} = C_{i0} \supseteq C_{i1} \supseteq \ldots \supseteq C_{is} = C_i;$$

on putting all these pieces together we obtain a refinement of (2.17):

$$M = C_{00} \supseteq C_{01} \supseteq \ldots \supseteq C_{0s} = C_{10} \supseteq \ldots \supseteq C_{r-1s} = C_{r0} \supseteq \ldots \supseteq C_{rs} = 0. \tag{2.19}$$

By symmetry the submodules

$$D_{ji} = (C_i \cap D_{j-1}) + D_j, \qquad i = 0, \ldots, r, \quad j = 1, \ldots, s,$$

provide a refinement of (2.18):

$$M = D_{00} \supseteq D_{01} \supseteq \ldots \supseteq D_{0r} = D_{10} \supseteq \ldots \supseteq D_{s-1r} = D_{s0} \supseteq \ldots \supseteq D_{sr} = 0. \tag{2.20}$$

By the Zassenhaus lemma we have $C_{ij-1}/C_{ij} \cong D_{ji-1}/D_{ji}$ and this shows that the chains (2.19) and (2.10) are isomorphic. Thus we have proved the **Schreier refinement theorem** (Otto Schreier, 1901–1929):

Theorem 2.8

In any module any two chains of submodules have isomorphic refinements. ■

We remark that any repetition in a chain corresponds to a trivial factor. Omitting such factors, we may assume that the refinements are without repetitions in Theorem 2.8. This theorem is of most interest when the chains are composition series, for then they have no proper refinements. Let M be a module with a composition series. This series has no proper refinements, so if we compare it with another chain, we find that the latter has a refinement with simple factors, i.e. a composition series. Thus we obtain the Jordan–Hölder theorem (Camille Jordan, 1838–1922; Otto Hölder, 1859–1937):

Theorem 2.9

Let M be a module with a composition series. Then any chain in M without repetitions can be refined to a composition series and any two composition series are isomorphic. In particular, all composition series of M have the same length. ■

To give an example, let m be a positive integer and take the ring $\mathbb{Z}/(m)$ of integers mod m, as $\mathbb{Z}$-module. If $m = p_1p_2\dots p_r$ is a complete factorization of m into primes, then

$$\mathbb{Z}/(m) \supseteq (p_1)/(m) \supseteq (p_1p_2)/(m) \supseteq \dots \supseteq 0$$

is a composition series; the Jordan–Hölder theorem tells us in this case that every positive integer has a complete factorization into prime numbers and in any two complete factorizations of m the number of factors is the same and the factors are the same up to unit factors. Here the last part follows because $\mathbb{Z}/(p) \cong \mathbb{Z}/(q)$ holds iff $p = qu$ for some unit u (which in $\mathbb{Z}$ means $u = \pm 1$). Thus Theorem 2.9 applied to $\mathbb{Z}$ tells us that $\mathbb{Z}$ is a unique factorization domain (cf. Section 3.3).

The following simple criterion allows us to decide which modules have a composition series.

Proposition 2.10

For any module M the following conditions are equivalent:

(a) M has a composition series;

(b) every chain in M without repetitions can be refined to a composition series;

(c) M is both Noetherian and Artinian.

Proof

(a) $\Rightarrow$ (b) follows from Theorem 2.9. To prove (b) $\Rightarrow$ (c), take any ascending chain in M; by hypothesis it can be refined to a composition series, which by definition is finite, hence the original chain was finite. This shows M to be Noetherian; similarly it follows that M is Artinian. (c) $\Rightarrow$ (a). Take a maximal submodule C_1 of M and generally if a chain $M = C_0 \supseteq C_1 \supseteq \ldots \supseteq C_r$ with simple factors C_i/C_{i+1} has been defined and if $C_r \neq 0$, take a maximal submodule C_{r+1} of C_r to continue. Since M is Artinian, this process must stop; when it does, we have found a composition series for M. ■

Exercises 2.2

1. Let k be a field and $R = k_n$ a full matrix ring over k. If C_i for $i = 0, 1, \ldots, n$, denotes the subspace of R in which the first i rows are zero, show that C_i is a right ideal of R and that $R = C_0 \supseteq C_1 \supseteq \ldots \supseteq C_n = 0$ is a composition series for R.

2. Show that for any submodules A, B of a module M we have $A + (B \cap A) = (A + B) \cap A = A$ (this is sometimes called the **absorptive law**).

3. Show that any injective endomorphism of an Artinian module is an automorphism. Show also that any surjective endomorphism of a Noetherian module is an automorphism, and give counter-examples for a module without chain condition.

4. Determine all integers n such that a cyclic group of order n has only a single composition series. Are there any other finite abelian groups with this property?

5. Given an exact sequence of modules, each of finite composition length, $0 \to M_1 \to M_2 \to \ldots \to M_r \to 0$, show that $\sum_i (-1)^i \ell(M_i) = 0$.

6. Let M be a module with a composition series. Show that every submodule of M is a member of a composition series for M.

7. Given an infinite descending chain of ideals in $\mathbb{Z}$, $C_1 \supset C_2 \supset \ldots$, show that $\bigcap C_i = 0$.

8. In a given module M with a submodule N, let $A_1 \subset A_2 \subset \ldots$ be an ascending chain of submodules with simple factors A_{i+1}/A_i. Show that $A_i + (N \cap A_{i+1}) = A_j$, where $j = i$ if $N \subseteq A_i$ and $j = i + 1$ if $A_{i+1} \subseteq N$.

9. Let M be any module with submodules A, B and C. Show that $(A+B) \cap C = A \cap C + B \cap C$ holds iff $(A \cap B) + C = (A+C) \cap (B+C)$ (when both of these relations are satisfied for all submodules, M is said to be **distributive**).

10. Show that a module M is distributive (cf. Exercise 9) iff for any homomorphism f from a module to M and any submodules A, B, $(A+B)f^{-1} = Af^{-1} + Bf^{-1}$, or equivalently, for any homomorphism g from M to a module and any submodules A, B, $(A \cap B)g = Ag \cap Bg$.

2.3 Artinian Rings: the Semisimple Case

A ring R is said to be **right Artinian** if it is Artinian when regarded as a right module over itself. This then means that R satisfies the minimum condition on right ideals. **Left Artinian** rings are defined similarly, and by an **Artinian** ring we understand a ring which is left and right Artinian. For example, a full matrix ring over a field k, $\mathfrak{M}_n(k)$, is Artinian. Similarly R is called **right Noetherian** if it is Noetherian when regarded as a right R-module, with corresponding definitions for (left) Noetherian. Later (in Section 2.4) we shall see that every Artinian ring is Noetherian; this will follow from the theory to be developed, but for the moment this result is not needed.

Let R be a right Artinian ring and A a right ideal of R. Then R/A is an Artinian R-module, because its submodules correspond to the right ideals of R containing A, by the third isomorphism theorem. It follows that every cyclic right R-module is Artinian, for any such module is of the form R/A, for some right ideal A. More generally this holds for any finitely generated right R-module, as the next result shows:

Theorem 2.11

Let R be a right Artinian ring. Then any finitely generated right R-module is again Artinian.

Proof

Let M be any finitely generated right R-module and take a generating set $u_1, \ldots, u_r$ say. We shall use induction on r to prove the result. When $r = 1$, M is cyclic and we have seen that the result then holds. Now let $r > 1$ and write M' for the submodule of M generated by $u_2, \ldots, u_r$. By induction M' is Artinian and M/M' is Artinian because it is cyclic, generated by the coset $u_1 + M'$. Now take any descending chain in M:

$$M = C_0 \supseteq C_1 \supseteq \ldots \supseteq C_r \supseteq \ldots.$$

The modules $C_i \cap M'$ form a descending chain in M' and so we have equality from some point onwards. Similarly the modules $C_i + M'$ form a descending chain from M to M', corresponding to a chain in M/M' and so we again have equality from some point onwards. Say for $n \geq r$ we have $C_n \cap M' = C_r \cap M'$, $C_n + M' = C_r + M'$; it follows that $C_n = C_r$ for all $n \geq r$, by Theorem 2.5, and this shows that every descending chain in M becomes stationary, i.e. M is Artinian, as claimed. ■

We shall now investigate the structure of a certain class of Artinian rings, but first we dispose of a special case.

Theorem 2.12

Any integral domain which is right Artinian is a skew field.

Proof

Let R be an integral domain which is right Artinian and take $a \in R$ such that a is neither zero nor a unit. Then $a^n \notin a^{n+1}R$, for if we had $a^n = a^{n+1}b$, say, then by cancellation $ab = 1$, hence $aba = a$ and so $ba = 1$, thus a is a unit and we have a contradiction. Hence $a^n \notin a^{n+1}R$, so $a^{n+1}R \subset a^n R$ and we have a strictly descending chain

$$R \supset aR \supset a^2R \supset \ldots,$$

but in a right Artinian ring every strictly descending chain must break off, so we have another contradiction and the conclusion follows. ■

This result shows in particular that any (non-zero) finite-dimensional algebra without zero-divisors must be a skew field. A skew field finite-dimensional over its centre is also called a **division algebra**; thus any finite-dimensional algebra A which is an integral domain is a division algebra, since its centre K is a subfield containing the ground field and A is again finite-dimensional over K.

In our study of Artinian rings we shall need a remark about modules which is proved quite simply but is of fundamental importance, **Schur's lemma** (after Issai Schur, 1875–1941, who introduced it in 1905 in his work on group characters).

Theorem 2.13

Let M be a simple module over a ring R. Then $\mathrm{End}_R(M)$ is a skew field. More generally, any homomorphism between two simple R-modules is either zero or an isomorphism.

Proof

Let $f : M \to N$ be a homomorphism between simple modules; if $f \neq 0$, then $\mathrm{im}\ f \neq 0$, hence $\mathrm{im}\ f = N$ and $\ker f \neq M$, therefore $\ker f = 0$. Thus f is one-one and onto, and so it is an isomorphism, and the second part is proved. Suppose now that $N = M$; the endomorphisms of M form a ring and by what has been shown, they are either zero or invertible, hence $\mathrm{End}_R(M)$ is a skew field, as claimed. ■

Suppose that A is a k-algebra, where k is an algebraically closed field, for example the complex numbers. Then as we have seen, every A-module is a k-space. In this case Theorem 2.13 has the following useful consequence.

Corollary 2.14

Let A be a k-algebra, where k is an algebraically closed field. Then for any simple A-module M, which is finite-dimensional as k-space, $\mathrm{End}_R(M) \cong k$.

Proof

Let α be an endomorphism of M. From Section 1.1 we know that α is a k-linear transformation of M, and by Theorem 2.13 it is either zero or invertible. This linear transformation has an eigenvalue λ, because k is algebraically closed. Thus $\alpha - \lambda.1$ is singular, and hence equal to zero, i.e. $\alpha = \lambda.1$; this shows $\mathrm{End}_R(M)$ to be one-dimensional. ■

We now come to introduce a class of rings for which there is a good structure theory. A module is said to be **semisimple** if it can be expressed as a direct sum of simple submodules; a ring R is called **semisimple** if it is semisimple as a right module over itself. At first sight it looks as if this should be called "right semisimple", but we shall soon see that right and left semisimple, for any ring, are equivalent. It also looks as if we might need infinite direct sums, but in fact every semisimple ring is a finite direct sum. We see this as follows: let R be any semisimple ring and form a direct sum of minimal right ideals: $B_r = A_1 \oplus A_2 \oplus \ldots \oplus A_r$. We thus have a strictly ascending chain of right ideals $B_1 \subset B_2 \subset \ldots$ whose union is R; but R is generated by 1, as right ideal, so if $1 \in B_n$, say, then $B_n = R$ and we see that R is a direct sum of a finite number of minimal right ideals. In particular, it follows that R is both Artinian and Noetherian. We state the result as

Theorem 2.15

Any semisimple ring R can be expressed as a direct sum of a finite number of minimal right ideals; in particular it is Artinian and Noetherian.

To prove the last part, we note that by the first part R has a composition series, so we can apply Proposition 2.10 to reach the conclusion. ■

Proposition 2.16

A finitely generated module M is semisimple if and only if it can be expressed as a sum of simple submodules. Moreover, any finitely generated module over a semisimple ring is semisimple.

The result actually holds for all modules, not necessarily finitely generated, but we have stated a special case which is simpler to prove and which is all we need here. To prove the result, assume that $M = \sum C_i$, where each C_i is simple. Since M is finitely generated, we can find a finite sum of simple submodules C_i which contains a generating set and so equals M. We define $N_1 = C_1$ and, generally, if N_j has been defined as a direct sum of some of the simple modules C_i for $j \leq r$, assume that $N_r \neq R$; then there is a C_k which is not contained in N_r. It follows that $N_r \cap C_k \neq C_k$ and therefore $N_r \cap C_k = 0$. Hence the sum $N_{r+1} = N_r \oplus C_k$ is direct. The process ends when N_n contains all the C_i and so equals M. Thus $M = N_n$ and this expresses M as a direct sum of simple submodules. For the second part we note that we have a surjection $f : R^n \to M$. Now R^n is a direct sum of simple modules, and under the homomorphism f each stays simple or becomes zero, so M is a sum of simple modules and so is semisimple. ■

To study the structure of semisimple rings we shall need a couple of general remarks. Given a ring R, as we saw in Section 2.1, we can with each element c of R associate two operations on R, the **right multiplication** $\rho_c : x \mapsto xc$ and the **left multiplication** $\lambda_c : x \mapsto cx$. If we regard R as a left R-module ${}_RR$ in the natural way, we see that each right multiplication is an endomorphism; this is just the associative law: $a(xc) = (ax)c$ holds for all $a \in R$. But the converse also holds; we have

Theorem 2.17

Let R be any ring. Then $\text{End}_R({}_RR)$, the ring of all endomorphisms of R, qua left R-module, is just the set of all right multiplications. Similarly $\text{End}_R(R_R)$ is the set of all left multiplications. More precisely, we have $\text{End}_R({}_RR) \cong R$, $\text{End}_R(R_R) \cong R^\circ$.

Proof

We have already seen that each right multiplication ρ_c defines an endomorphism of ${}_RR$, thus we have a mapping $\rho : R \to \text{End}_R({}_RR)$. We claim that this is a ring isomorphism. It is clearly linear and we have $\rho_a\rho_b : x \mapsto xa \mapsto xab$, hence $\rho_a\rho_b = \rho_{ab}$ and so ρ is a homomorphism. It is injective, because if $\rho_a = 0$, then $a = 1.a = 1\rho_a = 0$. Further, for any endomorphism φ of ${}_RR$, put $1\varphi = a$; then we have for any $x \in R$, $x\varphi = x(1\varphi) = xa = x\rho_a$, hence $\varphi = \rho_a$ and this shows ρ to be surjective. Thus the endomorphisms of ${}_RR$ form a ring isomorphic to R and by symmetry (or applying this result to the opposite ring R°), we find that the endomorphisms of R_R form a ring isomorphic to R°, i.e. anti-isomorphic to R. ■

Sometimes we have the situation ${}_RR_S$, where the ring R is regarded as an R-S-bimodule. Since S acts by R-endomorphisms, its action on R is represented by right multiplications; now the endomorphisms of R as bimodule are the right multiplications that commute with the actions of the elements of S, i.e. the centralizer of this action. Thus we obtain

Corollary 2.18

Let R be a ring which is an R-S-bimodule. Then the action of S is represented by right multiplication by elements of a subring S_0 of R and the endomorphism ring is the centralizer of S_0 in R. ■

Secondly we need an expression for the endomorphism ring of a direct sum of modules. For any ring R, consider a left R-module M which is expressed as a direct sum of a finite number of submodules:

$$M = C_1 \oplus \ldots \oplus C_n. \tag{2.21}$$

With each submodule C_i we can associate the canonical projection $\pi_i : M \to C_i$, given by $(x_1, \ldots, x_n) \mapsto x_i$ and the canonical inclusion $\mu_i : C_i \to M$, given by

$x \mapsto (0, \ldots, 0, x, 0, \ldots, 0)$ where x occupies the i-th place. These mappings satisfy the equations

$$\sum_i \pi_i \mu_i = 1 \tag{2.22}$$

and

$$\mu_i \pi_j = \delta_{ij}, \tag{2.23}$$

where δ_{ij} is the Kronecker delta defined in Section 1.2. Each endomorphism f of M gives rise to a matrix (f_{ij}), where $f_{ij} : C_i \to C_j$ is given by the equation $f_{ij} = \mu_i f \pi_j$, and conversely, any family (h_{ij}) of homomorphisms $h_{ij} : C_i \to C_j$ defines an endomorphism h of M by the rule $h = \sum \pi_i h_{ij} \mu_j$. These two processes are mutually inverse: given f, if $f_{ij} = \mu_i f \pi_j$, then $\sum \pi_i f_{ij} \mu_j = \sum \pi_i \mu_i f \pi_j \mu_j = f$, by (2.22); conversely, if $h = \sum \pi_i h_{ij} \mu_j$, then $\mu_r h \pi_s = \sum \mu_r \pi_i h_{ij} \mu_j \pi_s = h_{rs}$ by (2.23). Further, the mapping $f \mapsto (f_{ij})$ is an isomorphism, since $(f+g)_{ij} = \mu_i (f+g) \pi_j = \mu_i f \pi_j + \mu_i g \pi_j = f_{ij} + g_{ij}$ and $(fg)_{ik} = \mu_i f g \pi_k = \sum \mu_i f \pi_j \mu_j g \pi_k = \sum f_{ij} g_{jk}$. The result may be stated as

Theorem 2.19

Let R be any ring and let M be a left R-module, expressed as a direct sum of submodules:

$$M = C_1 \oplus C_2 \oplus \ldots \oplus C_n. \tag{2.24}$$

Then each endomorphism f of M can be expressed as a matrix (f_{ij}), where $f_{ij} : C_i \to C_j$, and the correspondence $f \mapsto (f_{ij})$ is an isomorphism. ■

Here it should be understood that the matrices of the form (f_{ij}), where f_{ij} is a homomorphism from C_i to C_j, form a ring under the usual matrix addition and multiplication, and this ring is isomorphic to $\mathrm{End}_R(M)$. The result takes on a particularly interesting form when all the C_i in (2.24) are isomorphic:

Corollary 2.20

Let R be a ring, C an R-module and $S = \mathrm{End}_R(C)$. Then we have, for any positive integer n,

$$\mathrm{End}_R(C^n) \cong \mathfrak{M}_n(S).$$

Proof

This is just the special case of Theorem 2.19 where all the C_i are isomorphic and so each f_{ij} is an endomorphism of C, i.e. an element of S. ■

We now return to consider semisimple rings. Let us first take the case of semisimple rings in which all the minimal right ideals are isomorphic; it turns out that these are just the **simple** Artinian rings, i.e. those without ideals other than R itself and zero. The results were first obtained by Joseph Henry Maclagan Wedderburn (1882–1948) for finite-dimensional algebras in 1908. Later, in 1928, E. Artin proved the results for rings satisfying both chain conditions, and in 1929 E. Noether observed that it was enough to assume the minimum condition.

Theorem 2.21 (Wedderburn's First Structure Theorem)

For any ring R the following conditions are equivalent:

(a) R is simple and right Artinian;

(b) R is semisimple non-zero and all simple right R-modules are isomorphic;

(c) $R \cong \mathfrak{M}_n(K)$, where K is a skew field and n is a positive integer;

(a°)–(c°) the left-hand analogues of (a)–(c).

Moreover, the integer n in (c) is unique and the skew field K is unique up to isomorphism.

Proof

(a) $\Rightarrow$ (b). Let aR be a minimal right ideal of R. Then $R = RaR = \sum c_i aR$ by the simplicity of R and each $c_i aR$ is a homomorphic image of aR, hence either 0 or isomorphic to aR; thus R has been expressed as a sum of pairwise isomorphic simple right R-modules. By Proposition 2.16 it can be expressed as a direct sum of certain of these modules, so we have $R = A^n$, where A is a minimal right ideal of R; thus (b) holds.

(b) $\Rightarrow$ (c). Assume that $R = A^n$, where A is a minimal right ideal. By Corollary 2.20 we have

$$\mathrm{End}_R(R_R) \cong \mathfrak{M}_n(\mathrm{End}_R(A)). \tag{2.25}$$

Now $\mathrm{End}_R(R_R) = R^\circ$, by Theorem 2.17, while $\mathrm{End}_R(A)$ is a skew field (Schur's lemma), so its opposite is also a skew field, K say and the isomorphism (2.25), or rather, its opposite, reads $R \cong \mathfrak{M}_n(K)$, as we had to prove. (c) $\Rightarrow$ (a). If $R = \mathfrak{M}_n(K)$, it is clear that R is right Artinian, since it has a composition series of length n. To prove simplicity, let $a = (a_{ij}) \in R$ and denote the matrix units by e_{ij}. If $a \neq 0$, say $a_{rs} \neq 0$, then $e_{ij} a_{rs} = e_{ir} a e_{sj}$; since a_{rs} has an inverse, it follows that the ideal generated by a contains all the e_{ij}, hence it contains $\sum e_{ii} = 1$ and so it must be the whole ring. This shows R to be simple, and it proves the equivalence of (a)–(c); now the symmetry of (c) shows these conditions to be equivalent to their left-hand analogues. Further, the integer n is unique as the composition length of R, while the isomorphism type of K is unique because all minimal right ideals of R are isomorphic as R-modules. ■

Consider now the case of a semisimple ring R. We can write it as a direct sum of minimal right ideals:

$$R = C_1 \oplus C_2 \oplus \ldots \oplus C_r, \tag{2.26}$$

where the C_i are arranged in groups of isomorphic ones, say the C_i for $i = 1, \ldots, s_1$ are isomorphic, likewise for $i = s_1 + 1, \ldots, s_1 + s_2$, etc., while the C_i in different groups are non-isomorphic. From Theorem 2.19 we know that every endomorphism f of R can be expressed as (f_{ij}), where $f_{ij} : C_i \to C_j$. By Theorem 2.13, $f_{ij} = 0$ unless i and j belong to the same group; thus if $r = s_1 + \ldots + s_t$ then the matrix (f_{ij}) is a diagonal sum of t matrices whose entries are endomorphisms of the various C_i. We thus find that

$$R = \mathfrak{M}_{s_1}(K_1) \oplus \ldots \oplus \mathfrak{M}_{s_t}(K_t), \tag{2.27}$$

where $K_1 = \mathrm{End}_R(C_1)^\circ$, and similarly K_j is the opposite of the endomorphism ring of the minimal right ideals in the j-th group. This shows us that the first s_1 summands in (2.26) form a two-sided ideal of R, a fact that also follows from Theorem 2.19, because left multiplication of any C_i by an element of R produces either 0 or a right ideal isomorphic to C_i. In (2.27) we have a representation of R as a direct sum of ideals; it so happens that these ideals are themselves rings, but their unit elements are different from the unit element of R, so they are not subrings of R and it does not make sense to call R a "direct sum of rings"; however, we shall see later (in Section 4.1) that (2.27) represents R as a direct product of rings. We sum up the result as

Theorem 2.22 (Wedderburn's Second Structure Theorem)

A ring R is semisimple if and only if it can be written as a direct sum of a finite number of ideals which are themselves simple rings (and so are full matrix rings over skew fields). ■

As a particular consequence we note the symmetry of this condition:

Corollary 2.23

A ring R is semisimple if and only if its opposite is semisimple. ■

Exercises 2.3

1. Let R be a simple Artinian ring, say $R = K_r$. Show that there is only one simple right R-module up to isomorphism, S say, and that every finitely generated right R-module M is a direct sum of copies of S. If $M \cong S^k$ say, show that k is uniquely determined by M. What is the condition on k for M to be free?

2. Given a semisimple ring R, show that $R = A_1 \oplus \ldots \oplus A_r$, where each A_i is an ideal which as ring is simple Artinian. How many right ideals B, up to isomorphism, are there such that a power of B has R as a direct summand, while no right ideal $C \subset B$ has this property?

3. Let R be a semisimple ring and A an ideal in R. Show that A is itself a ring and that its unit element lies in the centre of R.

4. Let R be a right Artinian ring. Show that R_n is right Artinian, for any $n \geq 1$.

5. For any ring R and any $n \geq 1$, show that the centre of R_n is isomorphic to the centre of R. In particular, if k is a commutative field, then the centre of k_n is k.

6. Let R be a semisimple ring. Show that R is simple iff the centre of R is a field.

7. For a semisimple ring show that the number of isomorphism types of simple modules is finite and equal to the number of non-isomorphic minimal ideals.

8. Let R be a ring and M a right R-module. Show that for any $n \geq 1$, M^n can be considered as a right R_n-module in a natural way. What is the condition on M for M^n to be a cyclic R_n-module?

9. In any ring R define the **socle** S as the sum of all minimal right ideals of R. Show that S is a two-sided ideal of R, which is semisimple as right R-module.

10. An element e of a ring R is called an **idempotent** if $e^2 = e$. If e is an idempotent not 0 or 1, show that $R = eR \oplus (1 - e)R$ is a direct sum decomposition of R into two right ideals. What is the condition for eR to be a two-sided ideal?

2.4 Artinian Rings: the Radical

When we come to deal with general Artinian rings, there is no such neat structure theory as in the semisimple case; in fact this is one of the objects of current research. But as a first step we can ask whether the semisimple case can help us here. One way would be to form sums of minimal right ideals. Any Artinian ring R certainly has minimal right ideals and we can form the sum of all such right ideals; by induction (as in the proof of Proposition 2.16) this can be written as a direct sum of minimal right ideals, and this direct sum is in fact finite, for an infinite direct sum would not possess the Artinian property. This direct sum is called the **socle** of R, written $\mathrm{soc}(R)$; since left multiplication by an element of R maps a minimal right ideal again to a minimal right ideal or to 0, it follows that $\mathrm{soc}(R)$ is a two-sided ideal of R. So we can form the quotient $R/\mathrm{soc}(R)$, but it turns out that nothing much can be said about this quotient beyond what we know about R itself.

For this reason we shall not pursue this line, but instead look at ideals A of R such that the quotient R/A is semisimple. The least such ideal is called the **radical** of R and is written $\mathrm{rad}(R)$. That such a least ideal exists in any Artinian ring is clear; in fact we shall see later (in Section 4.6) that it exists quite generally. For the moment we note the uniqueness properties of the radical, which will be of use to us later.

Theorem 2.24

Let R be an Artinian ring and denote the intersection of all maximal right ideals of R by N. Then (i) N is a two-sided ideal and R/N is semisimple, (ii) N is contained in any ideal A such that R/A is semisimple, (iii) N is the intersection of all maximal left ideals.

Proof

We shall begin by proving (ii). Let A be any right ideal such that R/A is semisimple, say

$$R/A \cong C_1 \oplus C_2 \oplus \ldots \oplus C_n, \tag{2.28}$$

where the C_i are simple and the sum on the right is finite, because R/A is finitely generated. It follows that $C_2 \oplus \ldots \oplus C_n$ is a maximal submodule of the right-hand side of (2.28), which by the isomorphism (2.28) corresponds to a maximal right ideal of R, say M_1. Similarly we can form M_i corresponding to the maximal submodule obtained by omitting C_i. Clearly we have $\bigcap M_i = A$ and it follows that $A \supseteq N$.

To prove (i) we shall find a series of maximal right ideals M_i such that

$$M_1 \supset M_1 \cap M_2 \supset M_1 \cap M_2 \cap M_3 \supset \dots. \qquad (2.29)$$

If $M_1, \dots, M_r$ have been found to form such a chain and $\bigcap M_i = N_r$, suppose that some maximal right ideal M does not contain N_r. Then $M \cap N_r \subset N_r$ and the chain (2.29) can be continued. Because R is Artinian, the chain must break off and if this happens at the n-th stage, we have $M \supseteq N_n$ for all maximal right ideals M, hence N_n is the intersection of all maximal right ideals, i.e. $N_n = N$. Moreover, N was formed as the intersection of the finitely many maximal right ideals $M_1, \dots, M_n$. Since $N \subseteq M_i$, we have a natural homomorphism $\varphi_i : R/N \to R/M_i$ $(i = 1, \dots, n)$, and putting all these mappings together, we obtain a homomorphism

$$\varphi : R/N \to R/M_1 \oplus \dots \oplus R/M_n. \qquad (2.30)$$

We observe that on both sides of (2.30) the modules have the same composition length n. Moreover, φ is injective, because for any $x \in R$, $x\varphi_i = 0 \Leftrightarrow x \in M_i$, hence $x\varphi = 0 \Leftrightarrow x \in \cap M_i = N$. It follows that φ is an isomorphism and this shows R/N to be semisimple.

Finally, to prove (iii), we observe that by (i) and (ii), N is the least ideal such that R/N is semisimple; but we know that being semisimple is left-right symmetric, hence it follows that N is also the intersection of all maximal left ideals, as claimed. ∎

The radical of an Artinian ring has a further property which does not hold in the general case; to explain it we need another definition. An element c of a ring is called **nilpotent** if $c^n = 0$ for some positive integer n, the least such n being the **index** of nilpotence. For example, a matrix unit such as e_{12} is nilpotent, or more generally, any matrix (a_{ij}), where $a_{ij} = 0$ for $i \geq j$. To define nilpotence for ideals, let us first define the **product** of two ideals, or more generally, of two additive subgroups of R, say A and B, as $AB = \{\sum a_i b_i \mid a_i \in A, b_i \in B\}$. If $B = A$, we shall also write A^2 instead of AA and by induction define $A^n = A^{n-1}A$. If $A^n = 0$ for some $n \in \mathbb{N}$, A is said to be **nilpotent**. It is clear that in a nilpotent ideal every element is nilpotent, but the converse will not generally hold, though it does hold in Artinian rings (cf. Exercises 2.4.1 and 2.4.2).

We now have the following description of the radical in an Artinian ring:

Theorem 2.25

Let R be an Artinian ring. Then the radical rad(R) is the sum of all nilpotent right ideals and is itself nilpotent.

Proof

If I is any nilpotent right ideal of R, then RI is a two-sided ideal and $RIRI\dots RI = RII\dots I = 0$ for enough factors, so the ideal generated by I is nilpotent, and it is enough to confine our attention to two-sided ideals. We note that a semisimple ring has no nilpotent ideals other than zero, since any non-zero ideal is a direct sum of ideals isomorphic to full matrix rings and so is not nilpotent. It follows that in a general Artinian ring R, every nilpotent ideal is contained in $\mathrm{rad}(R)$. To show that $\mathrm{rad}(R)$ is nilpotent, let us put $N = \mathrm{rad}(R)$. We have $N \supseteq N^2 \supseteq \dots$ and since R is Artinian, equality must hold at some point, say $N^r = N^{r+1} = \dots$. We write $C = N^r$; then $C^2 = C$ and we shall show that $C = 0$. Suppose that $C \neq 0$ and take a minimal right ideal A in C such that $AC \neq 0$; then $xC \neq 0$ for some $x \in A$ and $(xC)C = xC^2 = xC$, hence $xC = A$ because A was minimal. It follows that $x = xy$ for some $y \in C$ and so $x(1-y) = 0$. Now y is nilpotent, say $y^n = 0$, therefore $(1-y)(1+y+\dots+y^{n-1}) = 1-y^n = 1$ and we conclude that $x = 0$, a contradiction. Hence $C = 0$ and so $\mathrm{rad}(R)$ is nilpotent, as claimed. ■

The following consequence of Theorem 2.25 is often used and is known as **Nakayama's Lemma** (after Tadasi Nakayama, 1912–1964) (for the somewhat complicated history of this lemma see Nagata (1962), p. 212):

Corollary 2.26

Let R be an Artinian ring, M a finitely generated right R-module and N a submodule of M. If $N + M\,\mathrm{rad}(R) = M$, then $N = M$.

Proof

Take a generating set $u_1, \dots, u_n$ of M and write $u_j = y_j + \sum u_i a_{ij}$, where $y_j \in N$, $a_{ij} \in \mathrm{rad}(R)$. As a matrix equation this may be written $\boldsymbol{u} = \boldsymbol{y} + \boldsymbol{u}A$, where $\boldsymbol{u}$, $\boldsymbol{y}$ are row-vectors with components u_j, resp. y_j and $A = (a_{ij})$. Now $\mathrm{rad}(R)$ is nilpotent, by Theorem 2.25, of index r, say; this means that the product of any r entries of the matrix A is zero, hence $A^r = 0$; as in the proof of Theorem 2.25 it follows that $I - A$ is a unit and so we have $u = y(I-A)^{-1}$. Thus each u_i can be expressed in terms of the y's and so $M = N$. ■

We remark that with an appropriate definition of the radical this result still holds for general rings; we shall return to this topic in Section 4.4.

As another consequence of Theorem 2.25 we have the fact that every Artinian ring is Noetherian. This is **Hopkins' Theorem** (after Charles Hopkins,

1902–1939). The same result was obtained independently at about the same time, 1939, by Jacob Levitzki, 1904–1956, but, because of wartime conditions, only published in 1945; we first prove a result for modules.

Theorem 2.27

For any right Artinian ring R and any right R-module M the following conditions are equivalent:

(a) M is Artinian;

(b) M is Noetherian;

(c) M has a composition series;

(d) M is finitely generated.

Proof

Let us write $N = \text{rad}(R)$; we know that the quotient $S = R/N$ is semisimple and N is nilpotent, say $N^r = 0$. Now consider the descending chain

$$M \supseteq MN \supseteq MN^2 \supseteq \ldots \supseteq MN^r = 0. \tag{2.31}$$

Each factor $F_i = MN^{i-1}/MN^i$ is annihilated by N and so may be regarded as an S-module. As an S-module F_i is semisimple, by Proposition 2.16; if M is Artinian, then each F_i is Artinian as R-module and so also as S-module, but being semisimple, it actually has a composition series. Putting all these series together, we obtain a composition series for M (as R-module), and this shows that (a) $\Rightarrow$ (c). The same argument applies if M is Noetherian, so we obtain (b) $\Rightarrow$ (c). Now (c) $\Rightarrow$ (b), (d) are clear and finally (d) $\Rightarrow$ (a) follows by Theorem 2.11. ■

Now Hopkins' theorem is obtained by applying the result to R:

Corollary 2.28

Every right Artinian ring is right Noetherian. ■

We remark that a right Artinian ring need not be left Artinian, see Exercise 2.4.4.

Exercises 2.4

1. In any ring an ideal consisting of nilpotent elements is called a **nilideal**. Show that in any Artinian ring each nilideal is contained in the radical and hence is nilpotent.

2. Let R be the ring of all matrices with countably many rows and columns, but only finitely many elements outside the main diagonal different from zero, and let N be the set of all upper 0-triangular matrices, i.e. of the form $A = (a_{ij})$, where $a_{ij} = 0$ for $i \geq j$. Show that N is an ideal in R which is nil but not nilpotent.

3. Let R be an Artinian ring with radical N. Show that N can also be defined as the intersection of all maximal two-sided ideals of R.

4. Show that $\mathbb{R}$, as a left $\mathbb{Q}$-module, is neither Artinian nor Noetherian. Let R be the ring of all 2×2 matrices of the form $\begin{pmatrix} a & b \\ 0 & c \end{pmatrix}$, where $a \in \mathbb{Q}$, $b, c \in \mathbb{R}$. Show that each left (resp. right) submodule of ${}_{\mathbb{Q}}\mathbb{R}_{\mathbb{R}}$ corresponds to a left (resp. right) ideal of R; deduce that R is right but not left Artinian.

5. Show that in an Artinian ring R any minimal right ideal A is either nilpotent, in which case $A^2 = 0$, or it contains an idempotent e and then $A = eR$.

6. Let R be a finite-dimensional algebra over a field (associative, but not necessarily with unit element). If R is a direct sum of minimal right ideals that are all non-nilpotent, show that R has a unit element.

7. Let R be an Artinian ring with radical N. Given any idempotent e in R, show that R can be written as the direct sum $R = eRe \oplus eU \oplus Ve \oplus U \cap V$, where $U = \{x \in A \mid xe = 0\}$, $V = \{x \in A \mid ex = 0\}$. Show that eRe has the radical eNe and $U \cap V$ has the radical $N \cap U \cap V$ (the direct sum decomposition of R is called the **Peirce decomposition**, after Benjamin Peirce, 1809–1880).

8. A ring is called **(semi)primary** if it is Artinian and its residue class ring by the radical is (semi)simple. Show that a semiprimary ring R can be written as a direct sum of ideals which are primary as rings, plus a subspace of the radical of R.

9. Let R be an Artinian ring with radical N and let r be the least integer such that $N^r = 0$. By regarding the modules N^i/N^{i+1} for $i = 0, 1, \ldots, r-1$ as (R/N)-modules, show that they are semisimple.

10. Let R, N, r be as in Exercise 9 and define an ideal $S^n(R)$ of R recursively as follows: $S^1(R) = \text{soc}(R)$ is the socle of R; if $S^i(R)$ is defined, then $S^{i+1}(R)$ is an ideal containing $S^i(R)$ such that $S^{i+1}(R)/S^i(R) = \text{soc}(R/S^i(R))$. Show that $S^r(R) = R$. ($S^i(R)$ is called the **ascending Loewy series** of R (Alfred Loewy, 1873–1935).)

2.5 The Krull–Schmidt Theorem

As we have seen in Section 2.3, over a semisimple ring every finitely generated module is a direct sum of simple modules, i.e. semisimple as a module. In particular, every indecomposable module is simple, where M is called **indecomposable** if it is non-zero and cannot be written as a direct sum of two non-zero submodules. This is no longer true for modules over more general rings, but the module can always be written as a direct sum of indecomposable modules, and when the module has finite length these summands are determined up to isomorphism. This is the content of the Krull–Schmidt theorem, which will be proved in this section.

As a first result we shall describe how an endomorphism of a module Mcan be used to obtain a decomposition of M. This is accomplished by **Fitting's Lemma** (after Hans Fitting, 1906–1938):

Lemma 2.29

Let R be any ring and M a left R-module of finite length. For any endomorphism λ of M there exists a decomposition

$$M = M_0 \oplus M_1, \tag{2.32}$$

such that λ maps each of M_0, M_1 into itself and the restriction of λ to M_0 is nilpotent, while the restriction to M_1 is an automorphism.

Proof

We have $M \supseteq M\lambda \supseteq M\lambda^2 \supseteq \ldots$, and since M has finite length, there exists $r(\leq \ell(M))$ such that $M\lambda^r = M\lambda^{r+1}$. Similarly, since $0 \subseteq \ker\lambda \subseteq \ker\lambda^2 \subseteq \ldots$, we have $\ker\lambda^s = \ker\lambda^{s+1}$ for some s. We take $n = \max(r, s)$ and note that $M\lambda^k = M\lambda^n$, ker λ^k = ker λ^n for any $k \geq n$. We now put $M\lambda^n = M_1$, ker $\lambda^n = M_0$ and claim that the conclusion holds for these submodules. Clearly each of M_0, M_1 admits λ as an endomorphism, and $M_1\lambda^n = M\lambda^{2n} = M\lambda^n = M_1$. Hence λ is surjective on M_1 and so is an automorphism on M_1, while $\lambda^n = 0$ on M_0. Thus for any $x \in M, x\lambda^n = u\lambda^n$ for some $u \in M_1$; further, we have $(x-u)\lambda^n = 0$, so $x - u \in M_0$, and hence $x \in M_0 + M_1$. This proves that $M = M_0 + M_1$, and this sum is direct, for if $z \in M_0 \cap M_1$, then $z = t\lambda^n$ and so $t\lambda^{2n} = z\lambda^n = 0$, hence $t \in \ker \lambda^{2n} = \ker \lambda^n$ and $z = t\lambda^n = 0$. This shows the sum $M_0 + M_1$ to be direct and it establishes the decomposition (2.32). ■

Suppose that M in Lemma 2.29 is indecomposable; then either M_0 or M_1 must be zero. Defining a **local ring** as a ring in which the set of non-units is an ideal, we have

Corollary 2.30

Given any ring R, let M be an indecomposable R-module of finite length. Then the nilpotent endomorphisms of M form an ideal and every non-nilpotent endomorphism is an automorphism, hence $\mathrm{End}_R(M)$ is a local ring. ■

We shall need a slight generalization of this result. Thus let M, N be two non-zero R-modules. Given homomorphisms $\alpha : M \to N$, $\beta : N \to M$, suppose that $\alpha\beta$ is an automorphism of M, with inverse γ. Then α is injective: if $x\alpha = 0$, then $x\alpha\beta = 0$, hence $x = 0$; further, β is surjective, for if $z \in M$, then $z = z\gamma\alpha\beta$. Hence we have a short exact sequence

$$0 \to \ker \beta \xrightarrow{\subseteq} N \xrightarrow{\beta} M \to 0;$$

since $\gamma\alpha\beta = 1$, this sequence is split by $\gamma\alpha$. Now im $\gamma\alpha$ = im α and so

$$N = \mathrm{im}\ \alpha \oplus \ker \beta. \tag{2.33}$$

In particular, if N is indecomposable, it follows that ker $\beta = 0$, and α, β are both isomorphisms.

We remark that in any direct decomposition $M = M' \oplus M''$ we have the projections $\pi' : M \to M'$, $\pi'' : M \to M''$ and the canonical inclusions $\mu' : M' \to M$, $\mu'' : M'' \to M$. The mappings $e' = \pi'\mu'$ and $e'' = \pi''\mu''$ are idempotents in $\mathrm{End}_R(M)$ such that

$$e' + e'' = 1; \tag{2.34}$$

they are called the **projections** determined by the decomposition $M = M' \oplus M''$. Conversely, any decomposition (2.34) of 1 into idempotents in $\mathrm{End}_R(M)$ defines a direct decomposition, as is easily verified. Of course a similar decomposition of 1 into idempotents exists in the case of more than two direct summands.

We can now prove the **Krull–Schmidt Theorem** (after Wolfgang Krull, 1899–1971 and Otto Yulevich Schmidt, 1891–1956):

Theorem 2.31

Let R be any ring. Any R-module M of finite length has a finite direct decomposition

$$M = M_1 \oplus \ldots \oplus M_r, \tag{2.35}$$

where each M_i is indecomposable, and this decomposition is unique up to isomorphism and the order of the terms: if $M = N_1 \oplus \ldots \oplus N_s$ is a second direct decomposition of M into indecomposable summands, then $s = r$ and $M_i \cong N_{i'}$ for some permutation $i \mapsto i'$ of $1, \ldots, r$.

Proof

The existence of (2.35) is clear: since M has finite length, we can simply take a direct decomposition with as many terms as possible. To prove the uniqueness we shall use induction on r. When $r = 1$, M is indecomposable and there is nothing to prove, so assume that $r > 1$. Let $e_1, \ldots, e_r$ be the projections determined by the decomposition (2.35) and $f_1, \ldots, f_s$ the projections for the second decomposition. Consider $f_1 = \sum e_i f_1$; each $e_i f_1$, restricted to N_1 is an endomorphism of N_1, and by the indecomposability of N_1 is either nilpotent or an automorphism, but their sum is f_1, the identity on N_1, hence at least one is an automorphism of N_1. By renumbering the M_i we may take this to be $e_1 f_1$. Now $\theta = 1 - f_1(1 - e_1)$ is an automorphism of M, for if $x\theta = 0$, then $0 = x\theta f_1 = x f_1 e_1 f_1$, hence $x f_1 = 0$ and so $x = x\theta + x f_1(1 - e_1) = 0$; this shows θ to be injective and hence an automorphism of M. Moreover, if $x \in N_i$, then

$$x\theta = x f_i \theta = \begin{cases} x f_1 e_1 & \text{if } i = 1 \\ x & \text{if } i \geq 2 \end{cases}$$

Applying θ we find that $M = M_1 \oplus N_2 \oplus \ldots \oplus N_r$, hence

$$M/M_1 \cong M_2 \oplus \ldots \oplus M_r \cong N_2 \oplus \ldots \oplus N_s.$$

By induction we have $r - 1 = s - 1$, hence $r = s$, and after renumbering the M_i, we find that $M_i \cong N_i$, as we had to show. ∎

We note that this result applies to any finitely generated module over an Artinian ring, for by Theorem 2.27 such a module has finite length. Theorem 2.31 can be proved more generally for finite groups; in this form it was first stated by Wedderburn in 1909, with a partial proof, which was completed by Remak in 1911 (Robert Remak, 1888–1942?). The case of modules (i.e. abelian groups with operators) was treated by Krull in 1925 and that of general groups with operators by O. Yu. Schmidt in 1928. A further generalization was obtained in 1950 by Azumaya (Goro Azumaya, 1920–), who showed that if the components all have local endomorphism rings, the condition on the length (to be finite) can be omitted.

Exercises 2.5

1. Show that the set of all rational numbers with denominators prime to a given prime p is a local ring. Are there other local subrings in $\mathbb{Q}$?

2. Verify that for semisimple modules (of finite length) the Krull–Schmidt Theorem is a consequence of the Jordan–Hölder Theorem.

3. Let M be any R-module. Show that if $\mathrm{End}_R(M)$ is a local ring, then M is indecomposable.

4. Find all idempotents in $\mathbb{Z}/(120)$ and give a decomposition into indecomposable ideals.

5. Let R be an Artinian ring and M, N, N' any finitely generated R-modules such that $M \oplus N \cong M \oplus N'$. Show that $N \cong N'$.

6. Let R be an Artinian ring and M, N finitely generated R-modules such that $M^n \cong N^n$ for some $n \geq 1$. Show that $M \cong N$.

7. Given an R-module M, show that if $\mathrm{End}_R(M)$ contains no idempotent apart from 0 and 1, then M is indecomposable. Hence find an example of an indecomposable module whose endomorphism ring is not local (see Section 3.1, p. 107).

8. Let R be an Artinian ring. If $R = A_1 \oplus \ldots \oplus A_r$, where each A_i is an indecomposable ideal, show that every indecomposable ideal of R is one of the A_i.

9. Let R be a local ring and M an R-module. Show that the set of all submodules of M is totally ordered by inclusion iff every finitely generated submodule of M is cyclic or, equivalently, iff every 2-generator submodule is cyclic.

10. Let A be an $n \times n$ matrix over a field K. Show that there exists an invertible matrix P such that $P^{-1}AP$ is a diagonal sum of an invertible matrix and a nilpotent matrix. (Hint: use Fitting's Lemma.)

2.6 Group Representations. Definitions and General Properties

Groups are often defined abstractly, by prescribing generators and defining relations, and it is generally far from easy to study the properties of the group. The task is made easier if we have a concrete representation of the group, for example as a group of permutations, or as a group of rotations. A useful method, which can be applied to any group, is to represent it by matrices; this is a powerful method which has been used in physics whenever groups occur (e.g. quantum mechanics and relativity theory) and indeed in mathematics in the deeper study of groups.

Given a group G, a **representation** of G over a field k is a homomorphism

$$\rho : G \to \mathrm{GL}_r(k), \tag{2.36}$$

where $\mathrm{GL}_r(k)$, the **general linear group** of degree r, is the group of all invertible $r \times r$ matrices over k. Thus ρ associates with each $x \in G$ an invertible $r \times r$ matrix $\rho(x)$ over k such that

$$\rho(xy) = \rho(x)\rho(y) \quad \text{for all } x, y \in G. \tag{2.37}$$

Since $\rho(x)$ is invertible, we have $\rho(x^{-1}) = \rho(x)^{-1}$ and $\rho(1) = I$. The integer r is called the **degree** of the representation. Usually the group G is finite; the number of its elements, i.e. its **order**, will be denoted by $|G|$.

For example, the cyclic group of order 2, $C_2 = \{1, c\}$, has the representation $1 \mapsto 1$, $c \mapsto -1$; for order 3 we have $C_3 = \{1, c, c^2\}$, which is represented by $c \mapsto \omega = \sqrt[3]{1}$. But C_3 can also be represented over the real numbers; we need only find a matrix A such that $A^3 = I$, where $A \neq I$, to avoid trivialities. Such a matrix is given by $A = \begin{pmatrix} 0 & -1 \\ 1 & -1 \end{pmatrix}$; then ρ is defined by

$$\rho(c) = \begin{pmatrix} 0 & -1 \\ 1 & -1 \end{pmatrix}, \quad \rho(c^2) = \begin{pmatrix} -1 & 1 \\ -1 & 0 \end{pmatrix}, \quad \rho(1) = \begin{pmatrix} 1 & 0 \\ 0 & 1 \end{pmatrix}. \tag{2.38}$$

Of course even over $\mathbb{Q}$ we have a representation of degree 1 of C_3, obtained by mapping all elements to 1; this is the **trivial** representation. By contrast, a representation in which all group elements are represented by different matrices is said to be **faithful**, e.g. the representation (2.38) of C_3 is faithful.

Returning to the general case, if we have a representation ρ of a group G, this can be interpreted as an action on a vector space, as follows. Let V be a vector space of dimension r, where r is the degree of ρ, and choose a basis $u_1, \dots, u_r$ of V. We can define an action of G on V by the equations

$$u_i x = \sum \rho_{ij}(x) u_j \quad (x \in G). \tag{2.39}$$

By linearity this action extends to the whole of V. Explicitly, if $a = \sum \alpha_i u_i$, then $ax = \sum \alpha_i \rho_{ij}(x) u_j$; with the help of (2.36) it is easily verified that in this way V becomes a G-module. Conversely, if we are given a G-module which is a finite-dimensional vector space over k, then by choosing a basis of V and reading (2.39) in the other direction, we obtain a representation ρ of G of degree dim V. Here the G-module V with the basis $u_1, \dots, u_r$ is said to **afford** the representation ρ of G.

Let us see how the representation afforded by a G-module is affected by a change of basis. If we regard the basis $u_1, \dots, u_r$ as a column vector $\boldsymbol{u} = (u_1, \dots, u_r)^{\mathrm{T}}$, where T indicates transposition, we can write (2.39) more briefly as

$$\boldsymbol{u}x = \rho(x)\boldsymbol{u}. \tag{2.40}$$

Suppose now that $\boldsymbol{v} = (v_1, \dots, v_r)^{\mathrm{T}}$ is another column vector related to $\boldsymbol{u}$ by the equation $\boldsymbol{v} = P\boldsymbol{u}$. Then (2.40) in terms of $\boldsymbol{v}$ becomes

$$\boldsymbol{v}x = (P\boldsymbol{u})x = P(\boldsymbol{u}x) = P\rho(x)\boldsymbol{u} = P\rho(x)P^{-1}\boldsymbol{v}.$$

Hence if the representation afforded by V via the basis $\boldsymbol{v}$ is $\sigma(x)$, then

$$\sigma(x) = P\rho(x)P^{-1}. \tag{2.41}$$

Two representations ρ, σ related as in (2.41) are said to be **equivalent**, and what we have proved can be stated by saying that different bases of a G-module afford equivalent representations. Since P can be any invertible matrix, it follows that any equivalent representations of G are afforded by the same G-module for suitable bases. Moreover, two G-modules that afford the same representation must be isomorphic, for if U and V with bases $\boldsymbol{u}$, respectively, $\boldsymbol{v}$ afford the representation ρ of G, then we have $\boldsymbol{u}x = \rho(x)\boldsymbol{u}$, $\boldsymbol{v}x = \rho(x)\boldsymbol{v}$ and it follows that the mapping $\sum \alpha_i u_i \mapsto \sum \alpha_i v_i$ is an isomorphism of G-modules $U \to V$. We sum up the result as

Theorem 2.32

Let G be a group. Then any representation of G gives rise to a G-module and any G-module affords a representation of G; moreover, equivalent representations correspond to isomorphic G-modules. ■

Among the examples in Section 2.1 we met the group algebra kG, which for any group G and any field k was an algebra with the elements of G as basis and with multiplication determined by linearity from the multiplication in G. Let us see how this is related to the concepts just introduced.

Given a G-module V, we can define V as a kG-module by linearity, by putting for any $v \in V$ and $\sum a_x x \in kG$,

$$v\left(\sum a_x x\right) = \sum a_x (vx).$$

Here the summation is taken over all $x \in G$. We can thus apply all the concepts and results of module theory also to G-modules. For example, if V is a G-module, then its submodules are just the submodules of V as kG-module. If V' is a submodule of V, let us choose a basis of V adapted to this submodule; thus we have a basis $u_1, \ldots, u_r$ of V such that $u_1, \ldots, u_s$ is a basis of V'. If the action is given by (2.40), then we have $u_i x \in V'$ for $i \leq s$ and hence $\rho_{ij}(x) = 0$ for $i \leq s < j$. This shows ρ to be of the form

$$\rho(x) = \begin{pmatrix} \rho'(x) & 0 \\ \gamma(x) & \rho''(x) \end{pmatrix}. \tag{2.42}$$

We see that ρ' is a representation afforded by V', while ρ'' is afforded by the quotient module V/V' relative to the basis formed by the cosets of $u_{s+1}, \ldots, u_r$ mod V'. The representations ρ' and ρ'' are also referred to as **subrepresentations** of ρ.

A representation ρ is said to be **reducible** if it is equivalent to a representation of the form (2.42), where $0 < s < r$. In terms of modules this means that the corresponding G-module has a non-zero proper submodule, i.e. it is not simple. In the contrary case, when V is simple, the representation afforded by it is called **irreducible**.

Any finite-dimensional G-module V clearly has a composition series

$$V = V_0 \supset V_1 \supset V_2 \supset \ldots \supset V_d = 0,$$

where V_{i-1}/V_i is simple for $i = 1, \ldots, d$. By choosing a basis adapted to these submodules, we obtain a representation in the form

$$\rho(x) = \begin{pmatrix} \rho_1(x) & 0 & 0 & \ldots & 0 \\ * & \rho_2(x) & 0 & \ldots & 0 \\ * & * & & \ldots & \vdots \\ * & * & & \ldots & \rho_d(x) \end{pmatrix} \tag{2.43}$$

If we can make the transformation so that the starred part is zero and $\rho(x)$ becomes the diagonal sum of the $\rho_i(x)$, then ρ is said to be **completely reducible**. The corresponding module is then a direct sum of simple modules, i.e. it is semisimple.

So far all our G-modules have been right G-modules, but of course we also have **left** G-modules, defined by the rules $(xy)v = x(yv)$, $1v = v$. However, any left G-module may be regarded as a right G-module by defining $v \circ x = x^{-1}v$, for $x \in G, v \in V$. For we have $v \circ (xy) = (xy)^{-1}v = y^{-1}x^{-1}v = (v \circ x) \circ y$.

If we recall that for any ring a left R-module may be expressed as a right module over R°, the opposite ring, the above observation is explained by the fact that the group algebra kG has an anti-automorphism $*$, i.e. a linear mapping such that $(ab)^* = b^*a^*$; it is given by

$$\left(\sum a_x x\right)^* = \sum a_x x^{-1}. \tag{2.44}$$

We can now turn any left kG-module V into a right kG-module by defining $v \circ a = a^*v$ $(v \in V, a \in kG)$.

We note that the mapping $*$ defined by (2.44) satisfies $a^{**} = a$. Thus (2.44) shows that kG is an algebra with an involution, as defined in Section 1.1. Algebras with involution play an important role in the general theory of algebras (cf. Jacobson 1996).

From the above definitions it is clear that the simplest representations are the irreducible ones and, given a representation, it is natural to try and find an equivalent form which is in reduced form, and in this way, decompose it into its irreducible constituents. This task is made easier when we are dealing with completely reducible representations. We shall assume that our groups are finite and that the field is of characteristic zero, although most results still hold for general fields, as long as the characteristic is prime to the order of the group.

An important result states that over a field of characteristic zero every representation of a finite group is completely reducible. This is the content of **Maschke's Theorem** (after Heinrich Maschke, 1853–1908):

Theorem 2.33

Let G be a finite group and k a field of characteristic zero. Then every representation of G is completely reducible. Hence every G-module is semisimple.

Proof

Let ρ be a representation of G which is reduced, say

$$\rho(x) = \begin{pmatrix} \rho'(x) & 0 \\ \gamma(x) & \rho''(x) \end{pmatrix}. \tag{2.45}$$

Let ρ', ρ'' have degrees d', d'' respectively. To say that ρ is completely reducible amounts to finding an invertible matrix P such that $\rho(x)P = P(\rho'(x) \oplus \rho''(x))$, where $\oplus$ indicates the diagonal sum. Let us assume that P, like ρ, has triangular form, so that this equation becomes

$$\begin{pmatrix} \rho'(x) & 0 \\ \gamma(x) & \rho''(x) \end{pmatrix} \begin{pmatrix} I & 0 \\ C & I \end{pmatrix} = \begin{pmatrix} I & 0 \\ C & I \end{pmatrix} \begin{pmatrix} \rho'(x) & 0 \\ 0 & \rho''(x) \end{pmatrix}.$$

On multiplying out we find that only the (2,1)-block gives anything new:

$$\gamma(x) = C\rho'(x) - \rho''(x)C; \tag{2.46}$$

to complete the proof we need only find a matrix C to satisfy this equation. By substituting from (2.45) in the equation $\rho(xy) = \rho(x)\rho(y)$, we obtain

$$\gamma(xy) = \gamma(x)\rho'(y) + \rho''(x)\gamma(y), \tag{2.47}$$

which may be rewritten as

$$\gamma(x) = \gamma(x)\rho'(y)\rho'(y)^{-1} = [\gamma(xy) - \rho''(x)\gamma(y)]\rho'(y^{-1}).$$

Now G is of finite order m, say, and this is non-zero in k, by hypothesis. So by summing both sides as y runs over G, we obtain

$$m\gamma(x) = \sum_y [\gamma(xy) - \rho''(x)\gamma(y)]\rho'(y^{-1}).$$

In the first sum let us put $z = xy$; then $y^{-1} = z^{-1}x$, and when x is fixed, y runs over G as z does. So we have

$$\sum_y [\gamma(xy) - \rho''(x)\gamma(y)]\rho'(y)^{-1} = \sum_z \gamma(z)\rho'(z^{-1}x) - \sum_y \rho''(x)\gamma(y)\rho'(y^{-1}).$$

Since $\rho'(z^{-1}x) = \rho'(z^{-1})\rho'(x)$, the last two equations show that

$$m\gamma(x) = \sum_z \gamma(z)\rho'(z^{-1})\rho'(x) - \sum_y \rho''(x)\gamma(y)\rho'(y^{-1}).$$

This is just Eq. (2.46), if we write $m^{-1}\sum_y \gamma(y)\rho'(y^{-1})$ as C. Now the last part follows because as we noted in Section 2.5, semisimplicity corresponds to complete reducibility. ■

In the special case of a cyclic group of order n this result tells us that every matrix of finite order (over a field of characteristic zero) is similar to a diagonal matrix. This is of course a well-known result of linear algebra, which follows from the fact that the equation satisfied by the matrix, $x^n - 1 = 0$, has distinct roots.

For a further study of representations we shall use the group algebra kG; we remark that an element $f = \sum a(x)x$ of kG is essentially just a k-valued function on G. Our aim will be to show that the functions arising from inequivalent representations are orthogonal relative to a certain scalar product on kG. It is defined by setting

$$\left(\sum a(x)x, \sum b(y)y\right) = |G|^{-1}\sum a(x^{-1})b(x). \tag{2.48}$$

Clearly this form is bilinear and satisfies the equation

$$(g, f) = (f^*, g^*), \tag{2.49}$$

where * is the involution $\sum a(x)x \mapsto \sum a(x^{-1})x$ introduced earlier. Further, this form is **non-degenerate**, i.e. $(f, x) = 0$ for all x implies $f = 0$. For if $(f, x) = 0$, then $f(x^{-1}) = 0$, and when this holds for all $x \in G$, then $f = 0$.

We shall need a lemma on averaging the effect of a linear mapping:

Lemma 2.34

Let G be a group of finite order m and k a field of characteristic zero. Given two G-modules U, V and a linear mapping $\alpha : U \to V$, consider the mapping

$$\alpha^\bullet : u \mapsto m^{-1}\sum_x ((ux^{-1})\alpha)x. \tag{2.50}$$

This is a G-homomorphism from U to V. Moreover, for any G-homomorphism α, we have $\alpha^\bullet = \alpha$ and given mappings $\alpha : U \to V$, $\beta : V \to W$, if one of them is a G-homomorphism, then $(\alpha\beta)^\bullet = \alpha^\bullet\beta^\bullet$.

Proof

Fix $a \in G$ and write $y = xa$, $x = ya^{-1}$. Then as x runs over G, so does y and conversely. Given $\alpha : U \to V$, we have

$$m.u\alpha^\bullet a = \sum_x ux^{-1}\alpha xa = \sum_y uay^{-1}\alpha y = m.ua\alpha^\bullet. \tag{2.51}$$

Hence $a^\bullet$ is a G-homomorphism. Conversely, if α is a G-homomorphism, then each term in (2.51) reduces to $u\alpha a = ua\alpha$, hence $\alpha^\bullet = \alpha$. If further, $\beta : V \to W$ is given and α is a G-homomorphism, then $m.u(\alpha\beta)^\bullet = \sum_x ux^{-1}\alpha\beta x = \sum u\alpha x^{-1}\beta x = m.u\alpha^\bullet\beta^\bullet$, and similarly if β is a G-homomorphism. ■

We can now establish the orthogonality relations for representations.

Theorem 2.35

Let G be a finite group and k an algebraically closed field of characteristic zero. Any two irreducible representations ρ, σ of degrees c, d, respectively, satisfy the following relations:

$$\frac{1}{|G|}\sum_{x\in G}\rho_{ij}(x^{-1})\sigma_{kl}(x) = \begin{cases} 0 & \text{if } \rho \text{ and } \sigma \text{ are inequivalent,} \\ \dfrac{1}{d}\delta_{jk}\delta_{il} & \text{if } \rho = \sigma. \end{cases} \tag{2.52}$$

This relation expresses the fact that the different representation coefficients, regarded as components of a vector in the space kG, are orthogonal; however, nothing is asserted about the coefficients of two representations that are equivalent but different.

Proof

Let U, V with bases $u_1, \ldots, u_c$, $v_1, \ldots, v_d$ be spaces affording the representations ρ, σ respectively and define a linear mapping $\alpha_{jk} : U \to V$ by $u_i\alpha_{jk} = \delta_{ij}v_k$; thus the matrix for α_{jk} has as (i, l)-entry

$$(\alpha_{jk})_{il} = \delta_{ij}\delta_{kl}. \tag{2.53}$$

By Lemma 2.34 we can form $\alpha^{\bullet}_{jk}$, which is a G-homomorphism from U to V, whose matrix is obtained by averaging (2.53). Denoting the order of G by m, we obtain

$$\begin{aligned} m.(\alpha^{\bullet}_{jk})_{il} &= \sum_{h,r,x}\rho_{ih}(x^{-1})\delta_{hj}\delta_{kr}\sigma_{rl}(x) \\ &= \sum_{x}\rho_{ij}(x^{-1})\sigma_{kl}(x). \end{aligned}$$

If ρ, σ are inequivalent, then $\alpha^{\bullet}_{jk} = 0$ by Schur's lemma (Theorem 2.13), and this proves the first part of (2.52). Next put $\sigma = \rho$; then by Corollary 2.14, $\alpha^{\bullet}_{jk} = \lambda_{jk} \in k$, and so

$$\sum_{x}\rho_{ij}(x^{-1})\rho_{kl}(x) = m.\lambda_{jk}\delta_{li}. \tag{2.54}$$

To find the value of λ_{jk} we put $l = i$ and sum over i:

$$m.d.\lambda_{jk} = \sum_{i,x}\rho_{ki}(x)\rho_{ij}(x^{-1}) = \sum_{x}\rho_{kj}(1) = m.\delta_{jk};$$

dividing by md, we find that $\lambda_{jk} = d^{-1}\delta_{jk}$. If we insert this value in (2.54), we obtain the second part of (2.52), and the assertion is proved. ■

This theorem leads to a criterion for a representation to contain the trivial representation as a part:

Corollary 2.36

Let ρ be any representation of degree d of a finite group G. Then

$$\sum_x \rho_{ij}(x) = 0 \quad \text{for all } i, j = 1, \ldots, d, \tag{2.55}$$

if and only if ρ does not contain the trivial representation.

For a proof we suppose first that ρ is irreducible and non-trivial; if we apply (2.52) with σ the trivial representation, $\sigma(x) = 1$ for all $x \in G$, we obtain (2.55). In the general case we may take ρ to be reduced (by Theorem 2.33); it will then be a direct sum of irreducible representations, and clearly (2.55) holds iff none of these is the trivial representation. ■

In terms of the inner product (2.49) the relations (2.52) may be written as

$$(\rho_{ij}, \sigma_{kl}) = d^{-1}\delta_{jk}\delta_{il} \text{ or } 0.$$

It follows that the functions ρ_{ij} are linearly independent, in particular there cannot be more than dim $kG = m$ of them. Thus we have

Corollary 2.37

The coefficients of inequivalent irreducible representations of a finite group G are linearly independent. Hence the number of such representations is finite, and if their degrees are $d_1, \ldots, d_r$ then

$$\sum_{i=1}^{r} d_i^2 \leq |G|. \quad ■ \tag{2.56}$$

The results just found place an upper limit on the number of representations, but so far there is nothing to provide us with any representations at all. We shall now show how to find representations of G; in fact we shall show that by taking enough representations we can always achieve equality in (2.56). This means that every k-valued function on a finite group G can be written as a linear combination of coefficients of irreducible representations; this is expressed by saying that these coefficients form a **complete** system of functions on G.

In order to establish equality in (2.56) we take the regular representation of G, i.e. we take the group algebra kG as G-module under right multiplication. Its degree is m, the order of G, and it is completely reducible, by Theorem 2.33. Hence kG is semisimple as G-module and so kG is semisimple as algebra. Thus we can write it as a product of full matrix rings over skew fields; when k is algebraically closed, these fields reduce to k, by Schur's lemma in the form of Corollary 2.14. In this way we find

$$kG = \prod_{i=1}^{r} k_{d_i}. \tag{2.57}$$

The i-th factor on the right provides an irreducible representation of degree d_i of G, and these representations are inequivalent, because the product is direct, so that the coefficients corresponding to different factors are linearly independent. By counting dimensions on both sides of (2.57) we obtain the following relation:

$$\sum_{i=1}^{r} d_i^2 = |G|.$$

We claim that every irreducible representation of G is equivalent to some component of the regular representation. For if this were not the case, it would be orthogonal to the regular representation; to see that this cannot happen, let $\rho_{ij}(x)$ be the given representation and let $\tau_{uv}(x)$ be the regular representation; then $\tau_{uv}(x) = \delta_{xu,v}$, hence $\sum \rho_{ij}(x^{-1})\tau_{uv}(x) = \rho_{ij}(uv^{-1})$, and this cannot vanish for all values of i, j and of $u, v \in G$. Hence we have proved

Theorem 2.38

Let G be a finite group and k any algebraically closed field of characteristic zero. Then all irreducible representations are obtained by completely reducing the regular representation of G. If the inequivalent irreducible representations are $\rho_1, \ldots, \rho_r$ of degrees $d_1, \ldots, d_r$, then

$$\sum_{i=1}^{r} d_i^2 = |G|. \blacksquare \tag{2.58}$$

Group representations first appeared as one-dimensional representations of abelian groups in the number-theoretic studies of C.F. Gauss. The representation theory for finite groups was developed by Georg Frobenius (1849–1916) in the 1880s and 1890s using the group determinant, i.e. the determinant of the matrix of a general group element in the regular representation. The theory was greatly simplified by I. Schur in his dissertation of 1901, where he proved his lemma (Theorem 2.13).

Exercises 2.6

1. Let G be a finite group and ρ, σ two irreducible inequivalent representations of G of degrees r, s respectively. Show that for any $r \times s$ matrix A, $\sum_x \rho(x^{-1})A\sigma(x) = 0$.

2. Show that for any representation ρ of a finite group G, the representation ρ^* defined by $\rho^*(x) = \rho(x^{-1})^{\mathrm{T}}$ is again a representation of G (called the **contragredient** of ρ). Under what conditions on ρ is ρ^* equal to ρ?

3. Let G be a finite group and V a finite-dimensional complex vector space which is a G-module. Show that for any positive-definite hermitian form $h(u, v)$ on V (i.e. $h(v, u) = \overline{h(u, v)}$, $h(u, u) > 0$ unless $u = 0$), the form $(u, v) = \sum_{x \in G} h(ux, vx)$ is positive-definite hermitian and invariant under the action of G (i.e. $(ux, vx) = (u, v)$ for all $x \in G$).

4. Let G, V and (u, v) be as in Exercise 3. By choosing an orthonormal basis of V relative to the form (u, v) show that any representation of G is equivalent to a unitary representation, and deduce that V is semisimple (Maschke's Theorem).

5. Let δ be a representation of degree 1 of a group G. Show that for any representation ρ of G, $\rho\delta$ is again a representation of G, which is irreducible iff ρ is irreducible.

6. Let G be a finite group and V a finite-dimensional G-module over a field of characteristic zero. Show that if G acts trivially on every composition-factor of V, then the G-action on V is trivial.

7. Let k be a field of finite characteristic p and G a finite group of order divisible by p. Show that the group algebra kG contains a central nilpotent element and deduce that kG is not semisimple.

8. Let $S \neq I$ be a 2×2 matrix over $\mathbb{C}$ such that $S^2 = I$, $\det S = 1$. Show that $S = -I$. If S is as before, but $S^r = I$ for $r \geq 2$, show that S is similar to $\begin{pmatrix} \alpha & b \\ 0 & \alpha^{-1} \end{pmatrix}$, where $\alpha^r = 1$. Can b be made 0?

9. Show that for any integers $m, n, r > 1$ there is a group generated by two elements a, b such that a, b, ab have orders m, n, r, respectively, by finding a faithful representation in the projective special linear group $\mathrm{PSL}_2(\mathbb{C})$ (i.e. the quotient of $\mathrm{SL}_2(\mathbb{C})$, the special linear

group, consisting of matrices of determinant 1, by its centre). (Hint: take $\rho(a)$ upper and $\rho(b)$ lower triangular.)

10. Let G be a finite group and ρ any representation of G. Show that $N = \{x \in G \,|\, \det r(x) = 1\}$ is a normal subgroup. What can be said about G/N?

2.7 Group Characters

In the preceding section we have met group representations, which provide a concrete way of presenting groups, but even they are not easy to handle, since we have to deal either with modules or with matrix functions, which are still dependent on the choice of a basis. It will simplify matters to consider invariants such as the trace or determinant; in particular the trace, as a linear invariant of the representation, turns out to be especially useful. We therefore define, for any representation ρ of a finite group G, its **character** as

$$\chi(x) = \operatorname{tr} \rho(x), \quad \text{where } x \in G; \tag{2.59}$$

here tr denotes the trace of the matrix $\rho(x)$; thus $\operatorname{tr} \rho(x) = \sum_i \rho_{ii}(x)$, where $\rho(x) = (\rho_{ij}(x))$. The representation ρ is said to **afford** the character χ when (2.59) holds. Of course any representation of degree 1 is its own character; it is also called a **linear** character. In particular, every group G has the **trivial** or **principal** character $\chi(x) = 1$ for all $x \in G$. Throughout this section the ground field is taken to be $\mathbb{C}$, the complex numbers.

We note the following properties of characters.

Theorem 2.39

Let G be a finite group.

(i) The character of any representation of G is independent of the choice of basis, so equivalent representations afford the same character.

(ii) Any character χ is constant on conjugacy classes of G:

$$\chi(y^{-1}xy) = \chi(x) \quad \text{for all } x, y \in G. \tag{2.60}$$

(iii) For any $x \in G$ of order n, $\chi(x)$ is a sum of n-th roots of 1 and the degree of χ is $\chi(1)$.

(iv) If ρ', ρ'' are representations of G with characters χ', χ'', then the characters afforded by the direct sum $\rho' \oplus \rho''$ and tensor product $\rho' \otimes \rho''$ are $\chi' + \chi''$ and $\chi'\chi''$, respectively.

Proof

For the proof of (i) and (ii) we use the fact that $\mathrm{tr}(AB) = \mathrm{tr}(BA)$ for any matrices A, B of the same order. Let ρ be the given representation and χ its character; any representation equivalent to ρ is of the form $C^{-1}\rho(x)C$. Now (i) follows because $\mathrm{tr}(C^{-1}\rho(x)C) = \mathrm{tr}(\rho(x)CC^{-1}) = \mathrm{tr}(\rho(x))$, and the same argument proves (ii).

To prove (iii) we note that if $x^n = 1$, then $\rho(x)^r = 1$ for some r dividing n; by taking the least value of r we see that the minimal equation for $\rho(x)$ has distinct roots, hence $\rho(x)$ can be transformed to diagonal form and any diagonal element α satisfies $\alpha^r = 1$, hence $\alpha^n = 1$ and this shows α to be an n-th root of 1. When $x = 1$, then $\rho(x) = I$ and $\mathrm{tr}(I)$ is equal to the degree of ρ.

Finally, (iv) follows because $\mathrm{tr}(A \oplus B) = \mathrm{tr}\ A + \mathrm{tr}\ B$, and the tensor product $\rho' \otimes \rho''$ is defined in terms of modules U', U'' affording ρ', ρ'' as the tensor product $U' \otimes U''$ with the action of G given by $(u \otimes v)g = ug \otimes vg$. This means that the representation afforded is the tensor product of the matrices and now $\mathrm{tr}(A \otimes B) = \mathrm{tr}\ A.\mathrm{tr}\ B$. ■

The reader unfamiliar with tensor products is advised to turn to Section 4.3 for clarification of the last point.

For a closer study of characters we shall first take a brief look at a special case, the abelian groups. Let A be any abelian group, in additive notation. A character of A is just a homomorphism from A to the multiplicative group of complex numbers of absolute value 1. It will be more convenient to divide by $2\pi\mathrm{i}$ and take the logarithm; the effect will be that the values are real numbers between 0 and 1, more precisely, the value lies in the additive group of real numbers mod 1: $\mathbb{T} = \mathbb{R}/\mathbb{Z}$. These characters can be added and again form a group which is called the **dual** or **character group** of G and is denoted by $\mathrm{Hom}(A, \mathbb{T})$ or $A^\wedge$. For example, if $A = C_n$ is the cyclic group of order n, with generator γ, then $r\gamma = 0$ iff $n|r$; given a character χ of C_n, $x = \chi(\alpha)$ satisfies $nx = 0$ and moreover, every function $\chi(r\alpha)$ $r = 0, 1, \ldots, n-1$, is a character of C_n. By taking $\chi(\alpha)$ to be $1/n$, we obtain a generator for the dual group, clearly of order n, so we see that $C_n^\wedge$ is again cyclic of order n. It turns out that this result holds quite generally:

Theorem 2.40

Every finite abelian group is isomorphic to its dual:

$$A^\wedge \cong A. \tag{2.61}$$

Proof

We have just seen that this holds for the cyclic case. In general we have, by the basis theorem for finite abelian groups, $A = A_1 \oplus \ldots \oplus A_r$, where each A_i is cyclic. Hence

$$\begin{aligned}\mathrm{Hom}(A, \mathbb{T}) &\cong \mathrm{Hom}(A_1 \oplus \ldots \oplus A_r, \mathbb{T}) \\ &\cong \mathrm{Hom}(A_1, \mathbb{T}) \oplus \ldots \oplus \mathrm{Hom}(A_r, \mathbb{T}),\end{aligned}$$

and this is isomorphic to A, by what has been proved in the cyclic case. ■

It is important to note that this isomorphism depends on the choice of the decomposition of A; in fact there is no natural transformation from A to $A^\wedge$. However, there is such a natural transformation

$$A = f^A : A \to A^{\wedge\wedge}, \tag{2.62}$$

from A to its **bidual**, i.e. the second dual. To define it, we consider, for each $x \in A$, the mapping

$$f_x : \chi \mapsto \chi(x). \tag{2.63}$$

This is easily seen to be a mapping from $A^\wedge$ to $\mathbb{T}$, hence an element of $A^{\wedge\wedge}$. Further it is clear that the mapping (2.62) so defined is a homomorphism. We claim that it is a natural transformation. Thus let $h : A \to B$ be a homomorphism of finite abelian groups; it induces homomorphisms $f^\wedge : B^\wedge \to A^\wedge$ and $f^{\wedge\wedge} : A^{\wedge\wedge} \to B^{\wedge\wedge}$ and this yields the diagram

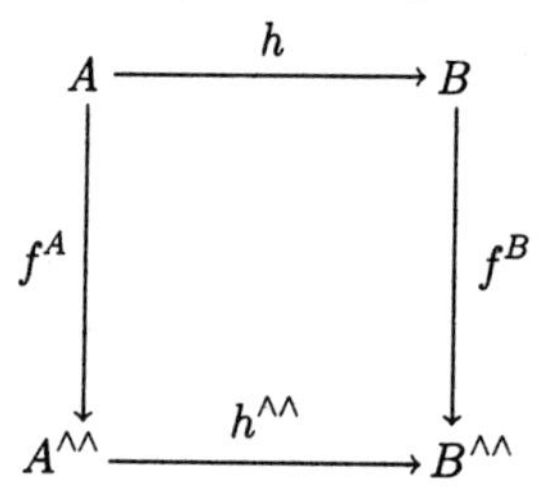

The mapping $h^{\wedge\wedge} : A^{\wedge\wedge} \to B^{\wedge\wedge}$ is given by $f^A h^{\wedge\wedge} = h f^B$ and this condition shows that the diagram above is commutative.

To find ker f we note that $f_x = 0$ means: $\chi(x) = 0$ for all $\chi \in A^\wedge$. But whenever $x \neq 0$, there is a character on the subgroup generated by x which does not vanish on x, and it can be extended to a character χ, say, on A, because $\mathbb{T}$ is divisible. Thus we have $\chi \in A^\wedge$ with $\chi(x) \neq 0$; hence $f_x \neq 0$ for $x \neq 0$ and this shows (2.62) to be injective. But A and $A^\wedge$ have the same order, hence so do $A^\wedge$ and $A^{\wedge\wedge}$, and it follows that f is an isomorphism. Thus we obtain

Theorem 2.41

For any finite abelian group A, the dual $A^\wedge$ is again finite and there is a natural isomorphism $A \cong A^{\wedge\wedge}$. ∎

This result has an interesting generalization to arbitrary finite groups:

Theorem 2.42

For any finite group G the number of linear characters is equal to the index $(G : G')$, where G' is the derived group. Hence every non-abelian group has irreducible representations of degree greater than 1.

Proof

Every homomorphism $f : G \to \mathbb{T}$ corresponds to a homomorphism from the abelian group $G/G'$and conversely, every homomorphism $G/G' \to \mathbb{T}$ corresponds to one from G to $\mathbb{T}$. By Theorem 2.40 the number of these homomorphisms is $|G/G'| = (G : G')$. Now the rest follows because for an abelian group G every irreducible character over $\mathbb{C}$ is linear, and so their number is $|G|$, by (2.58) of Section 2.6. ∎

We now return to consider the inner product defined on the group algebra in (2.48) of Section 2.6. For characters α, β this takes the form

$$(\alpha, \beta) = |G|^{-1} \sum \alpha(x^{-1})\beta(x). \tag{2.64}$$

Since every character α is a sum of roots of 1, $\alpha(x^{-1})$ and $\alpha(x)$ are complex conjugates and this shows (2.64) to be a hermitian scalar product. Let us see how the orthogonality relations of Theorem 2.35 translate to characters. If in the result we put $j = i$, $l = k$ and sum over i and k, then on writing the characters of ρ and σ as χ, ψ respectively, and remembering that equivalent representations afford equal characters, we obtain

$$(\chi, \psi) = \begin{cases} 1 & \text{if } \chi = \psi, \\ 0 & \text{otherwise.} \end{cases} \tag{2.65}$$

We see that under the metric defined by (2.64) the irreducible characters form an orthonormal system. If the different irreducible representations of G (up to equivalence) are $\rho_1, \dots, \rho_r$, with characters $\chi_1, \dots, \chi_r$, then any representation of G is, by complete reducibility (Theorem 2.33) equivalent to an expression

$$\rho = \nu_1\rho_1 \oplus \dots \oplus \nu_r\rho_r.$$

Its character, χ say, is obtained by taking traces on both sides:

$$\chi = \nu_1\chi_1 + \ldots + \nu_r\chi_r;$$

to find the value of ν_i we take scalar products with χ_i, using (2.65):

$$\nu_i = (\chi, \chi_i).$$

This shows in particular that over $\mathbb{C}$ any representation is determined up to equivalence by its character.

For another way of expressing the inner product (2.64) we use the modules affording the representations:

Proposition 2.43

Let G be a finite group and U, V any G-modules over $\mathbb{C}$, affording representations with characters α, β respectively. Then their inner product is given by

$$(\alpha, \beta) = \dim(\mathrm{Hom}_G(U, V)). \tag{2.66}$$

Proof

When U, V are simple G-modules, this follows from Schur's lemma (Theorem 2.13 and Corollary 2.14), for the dimension on the right is 1 or 0 according as V is or is not isomorphic to U. Now the general case follows by summation, using the semisimplicity of U and V. ■

The number on the right of (2.66) is often called the **intertwining number** of U and V. From the above proof we see that it is symmetric in $U$and V.

As we shall see, characters, being scalar functions, are easy to manipulate, but the information they yield is limited. In particular, though defined as functions on a group G, they are really class functions, i.e. they are constant on each conjugacy class and so they should be regarded as functions on the set of all conjugacy classes of G. By interpreting the orthogonality relations in this way we shall be able to show that the number of irreducible characters of G is equal to the number of its conjugacy classes.

Let us consider first a general function on $G : a(x)$ $(x \in G)$ and the corresponding element of the group algebra $kG : a = \sum a(x)x$, and find the condition for $a(x)$ to be a class function. We have

$$y^{-1}ay = \sum_y a(x)y^{-1}xy = \sum_z a(yzy^{-1})z \quad \text{for all } y \in G.$$

It follows that

$$y^{-1}ay = a \quad \text{for all } y \in G \quad \Leftrightarrow \quad a(yzy^{-1}) = a(z) \quad \text{for all } y, z \in G. \tag{2.67}$$

This shows that $a = \sum a(x)x$ lies in the centre of kG iff $a(x)$ is constant on the conjugacy classes. We can now state the result linking characters and class functions.

Theorem 2.44

Let G be a finite group and k an algebraically closed field of characteristic 0. Then a function $a(x)$ on G is a class function if and only if the corresponding element $a = \sum a(x)x$ of the group algebra kG lies in the centre. In particular, the class sums c_λ of elements in a given conjugacy class lie in the centre; in fact they form a k-basis for the centre and the irreducible characters over k form a basis for the class functions. Thus if the different irreducible characters of G are $\chi_1, \ldots, \chi_r$, then any class function φ on G has the form

$$\varphi = \sum_i (\chi_i, \varphi)\chi_i. \tag{2.68}$$

Hence the number of irreducible characters is equal to the number of conjugacy classes of G.

Proof

Let C_λ be a conjugacy class of G and denote the sum of the elements in C_λ by c_λ. The corresponding function is 1 on C_λ and 0 elsewhere, so is constant on the conjugacy classes; hence the elements c_λ lie in the centre of kG. Clearly every class function $a = \sum a(x)x$ can be expressed as a sum $\sum a_\lambda c_\lambda$ in just one way; hence the c_λ form a basis for the centre of kG.

Now let r be the number of irreducible characters and s the number of conjugacy classes, which we have seen is also the dimension of the centre of kG. The algebra kG is semisimple and so is a direct product of r full matrix rings over k. Each has a one-dimensional centre, so the centre of kG is r-dimensional, and it follows that $r = s$. By the orthogonality relation (2.65) the different characters are linearly independent, hence they form a basis; by (2.65) this basis is orthonormal, and so (2.68) follows. ∎

As an example let us work out the character table for the symmetric group S_3. This group, of order 6, has three classes: $C_1 = \{1\}$, $C_2 = \{(12), (13), (23)\}$, $C_3 = \{(123), (132)\}$. Hence there are three irreducible characters; if their degrees are d, d', d'', then the sum of their squares equals the order of the group, $d^2 + d'^2 + d''^2 = 6$. There are two one-dimensional representations, the trivial one and the sign representation, thus $d = d' = 1$ and so $d'' = 2$. If the character of this representation on the classes has values h_1, h_2, h_3, then

$h_1 = 2$, while the remaining values can be found from the orthogonality relations: we have $2 + 3h_2 + 2h_3 = 0$, $2 - 3h_2 + 2h_3 = 0$, whence $h_2 = 0$, $h_3 = -1$.

Let us write $\chi_{i\lambda}$ for the value of the character χ_i on the class C_λ and put $\delta_\lambda = |C_\lambda|^{1/2}$. Then the orthogonality relations can be written

$$|G|^{-1}\sum \delta_\lambda \bar{\chi}_{i\lambda} . \delta_\lambda \chi_{j\lambda} = \delta_{ij}.$$

This states that the matrix $S = (|G|^{-1/2}\delta_\lambda \chi_{i\lambda})$ is **unitary**, i.e. $S^{\mathrm{H}}S = I$, where S^{H} is the conjugate transpose of the matrix S. It follows that $SS^{\mathrm{H}} = I$, and hence

$$|G|^{-1}\sum \delta_\lambda \bar{\chi}_{i\lambda}\delta_\mu \chi_{i\mu} = \delta_{\lambda\mu}.$$

Here we can omit $\delta_\lambda\delta_\mu$ when $\lambda \neq \mu$; the result is a second orthogonality relation:

Theorem 2.45

Let G be a finite group and denote by $\chi_{i\lambda}$ the value of the i-th irreducible character on the class C_λ. Then

$$\sum_i \bar{\chi}_{i\lambda}\chi_{i\mu} = \begin{cases} \dfrac{|G|}{|C_\lambda|} & \text{if } \lambda = \mu, \\ 0 & \text{if } \lambda \neq \mu. \end{cases}$$

■

Exercises 2.7

1. Find the character table for the quaternion group (generated by i, j with $\mathrm{i}^2 = \mathrm{j}^2 = -1$, $\mathrm{ij} = -\mathrm{ji}$).
2. Find the character table for the alternating group of degree 4 (i.e. the group of all even permutations on four symbols).
3. Let A be a finite abelian group. Show that any irreducible representation of A over $\mathbb{C}$ is of degree 1. (Hint: use the fact that two commuting matrices over $\mathbb{C}$ can be transformed simultaneously to diagonal form.) Deduce that every complex representation is a direct sum of representations of degree 1.
4. Let χ be any character of a finite group G. Show that (χ, χ), defined as in (2.64), is a positive integer.
5. Let A be a finite abelian group and $A^\wedge$ its dual. Show that for any subgroup B of A the annihilator of B in $A^\wedge$ is a subgroup of $A^\wedge$. What is its order?

6. Show that if x, y are elements of a finite group that are not conjugate, then there exists a character α such that $\alpha(x) \neq \alpha(y)$.

7. Let G be a finite group and ρ a representation of G whose character vanishes on all elements $\neq 1$ of G. Show that ρ is a multiple of the regular representation.

8. Let G be a finite group, $A = kG$ its group algebra (over a field k of characteristic zero) and $I = eA$ a right ideal in A, with idempotent generator $e = \sum e(x)x$. Show that the character afforded by I is χ, given by $\chi(g) = \sum_x e(xg^{-1}x^{-1})$.

9. Let G be a finite group and in the group algebra kG define c_λ as the sum of elements in a conjugacy class C_λ. Verify that $c_1, \ldots, c_r$ (where r is the number of conjugacy classes) is a basis for the centre of kG. Deduce the relation $c_\lambda c_\mu = \sum_v \gamma_{\lambda\mu\nu} c_\nu$, where $\gamma_{\lambda\mu\nu}$ is a non-negative integer.

10. With the notation of the last exercise, let ρ be an irreducible representation of G. Show that $\rho(c_\lambda) = \eta_\lambda I$, where $\eta_\lambda \in k$. If ρ is of degree d and has the character χ, show that for any $x \in C_\lambda, \chi(x) = d\eta_\lambda / h_\lambda$, where $h_\lambda = |C_\lambda|$. Deduce that $\eta_\lambda \eta_\mu = \sum \gamma_{\lambda\mu\nu} \eta_\nu$ and hence show that η_μ is a root of $\det(xI - \Gamma_\mu) = 0$, where $\Gamma_\mu = (\gamma_{\lambda\mu\nu})$.

3
Noetherian Rings

Throughout mathematics there are many examples of Noetherian rings, starting with the integers, and in this chapter we shall describe some of the most important classes, polynomial rings and rings of algebraic integers, as well as some of their properties, the Euclidean algorithm and unique factorization.

3.1 Polynomial Rings

In Section 2.3 we briefly met Noetherian rings; we recall that a ring is right Noetherian if every right ideal is finitely generated. As a first consequence we have an analogue of Theorem 2.11.

Theorem 3.1

Let R be a right Noetherian ring. Then every finitely generated right R-module is again Noetherian.

The proof is as for Theorem 2.11; instead of a descending chain we now have an ascending chain and we can draw the same conclusions as before. ■

Many readers will have met polynomial functions before; they are functions of the form

$$f = a_0 + a_1x + a_2x^2 + \ldots + a_nx^n, \tag{3.1}$$

where the a_i are members of the ground field, e.g. the real numbers. Here x may be a variable, and for each real value of x, f is again a real number, or we may regard x as an "indeterminate", allowing us to handle expressions of the form (3.1). We shall use this second interpretation; thus our aim will be to show that for any ring R we can form the ring of all polynomials in x with coefficients in R. The elements of this ring are formal expressions (3.1), where $a_i \in R$; two such expressions, f, given by (3.1) and

$$g = b_0 + b_1x + \ldots + b_mx^m \tag{3.2}$$

are equal iff $a_0 = b_0$, $a_1 = b_1, \ldots$; we do not insist that $m = n$ for the equality $f = g$ to hold, but if $m < n$, say, then $a_{m+1} = \ldots = a_n = 0$. Polynomials are added componentwise; if $m \leq n$, say, then

$$f + g = a_0 + b_0 + (a_1 + b_1)x + \ldots + (a_m + b_m)x^m + \ldots + a_nx^n, \tag{3.3}$$

while multiplication is by the rule $ax^r bx^s = abx^{r+s}$, together with distributivity; thus

$$fg = a_0b_0 + (a_0b_1 + a_1b_0)x + (a_0b_2 + a_1b_1 + a_2b_0)x^2 + \ldots + a_nb_mx^{m+n}. \tag{3.4}$$

In this way we obtain a ring, the **polynomial ring** in x over R, usually denoted by $R[x]$. If f is given by (3.1) and $a_n \neq 0$, then f is said to have **degree** n and we write $\deg(f) = n$ and call a_nx^n the **leading term**, while a_n is called the **leading coefficient** of f. The polynomials of degree 0 are just the non-zero elements of R, also called the **constant** polynomials. Together with 0 they form a ring isomorphic to R, so R is embedded as a subring in $R[x]$. These results may be stated as follows:

Theorem 3.2

Given any ring R, there exists a polynomial ring $R[x]$ in an indeterminate x over R. Its members are the expressions (3.1) with addition defined by (3.3) and multiplication defined by (3.4). Moreover, R is embedded in $R[x]$ as the ring of constant polynomials.

Here x can be any symbol, though we should make sure that it is not the name of an element of R, to avoid confusion. Strictly speaking we also need to prove that such a ring exists. At first sight this may seem obvious, but what we need to show is that distinct polynomials really represent distinct elements. To do this, consider the set of all sequences

$$(a_0, a_1, a_2, \ldots), \quad (a_i \in R) \tag{3.5}$$

where $a_i = 0$ for all i greater than some integer. We define addition and multiplication by the rules

$$(a_0, a_1, \ldots) + (b_0, b_1, \ldots) = (a_0 + b_0, a_1 + b_1, \ldots),$$
$$(a_0, a_1, \ldots)(b_0, b_1, \ldots) = (a_0 b_0, a_0 b_1 + a_1 b_0, a_0 b_2 + a_1 b_1 + a_2 b_0, \ldots).$$

It is easily verified that the set of all these sequences forms a ring with the unit element $(1, 0, \ldots)$. Moreover, the elements of the form $(a, 0, 0, \ldots)$ form a subring isomorphic to R and the whole ring is generated over R by the element $x = (0, 1, 0, \ldots)$. Now it is easily verified that two polynomials are equal iff the coefficients of corresponding powers of x are equal. It will be enough to show that f, given by (3.1), is zero iff all the a_i are zero; this follows because if f is given by (3.1), then

$$1.f = (1, 0, \ldots)(a_0, a_1, \ldots) = (a_0, a_1, \ldots).$$

Thus if some a_i is different from zero, then $1.f \neq 0$ and so $f \neq 0$, as we wished to show. ■

This result shows that it is important to distinguish polynomials from polynomial functions. To illustrate the point, take the field $F = \mathbb{Z}/(2)$, consisting of two elements; every element of this field satisfies the equation $x^2 = x$; hence the polynomial $x^2 - x$ defines the zero function, although as a polynomial it is different from zero.

The process of forming $R[x]$ is sometimes called "adjoining an indeterminate x to the ring R". Of course this process can be repeated, to yield a polynomial ring in several indeterminates; we need only be careful to give different names to the different variables, thus we can form $R[x, y]$ or with r indeterminates, $R[x_1, \ldots, x_r]$.

The formula (3.4) for the product shows that the leading term of a product fg is the product of the leading terms; it follows that when R is an integral domain, the degree of fg is the sum of the degrees of f and g:

$$\deg(fg) = \deg(f) + \deg(g). \tag{3.6}$$

Here we need to assume that $f, g \neq 0$, for the equation to have a sense, since $\deg(0)$ has not yet been defined. If for example, $g = 0$, (3.6) reduces to $\deg(0) = \deg(f) + \deg(0)$; in order for this equation to hold for all f, at least formally, one usually defines the degree of 0 to be $-\infty$.

Sometimes we shall impose a condition on x, such as $x^n = a$. The resulting ring will be denoted by $R[x \mid x^n = a]$; it may be thought of as the residue-class ring of $R[x]$ by the ideal $(x^n - a)$, or also as the ring generated by x over R with the defining relation $x^n = a$. A similar notation is used when there is more than one generator and defining relation.

In Section 3.2 we shall study the special properties of a polynomial ring over a field; for the moment we note a property of polynomial rings which applies quite generally, the **Hilbert basis theorem** (after David Hilbert, 1862–1943):

Theorem 3.3

If R is any right Noetherian ring, the polynomial ring $R[x]$ is again right Noetherian.

The following short proof is due to Heidrun Sarges. Let us put $S = R[x]$; we assume that S is not right Noetherian and prove that the same must hold for R. Thus we can find a right ideal A in S which is not finitely generated. We shall form a sequence $f_1, f_2, \dots$ of polynomials in A with the property that f_{r+1} does not belong to the right ideal generated by $f_1, \dots, f_r$. That there is such a sequence follows because A is not finitely generated. In detail, we take f_1 to be a polynomial of least degree in A and if $f_1, \dots, f_r$ have been found in A, we take f_{r+1} to be a polynomial of least degree which belongs to A but not to $f_1S + \ldots + f_rS$. Since A cannot be finitely generated, this process can be continued indefinitely. We denote the degree of f_i by n_i and its leading coefficient by a_i. From the choice of n_i it follows that $n_1 \leq n_2 \leq \dots$. We claim that the strictly ascending chain

$$a_1R \subset a_1R + a_2R \subset \dots \tag{3.7}$$

does not break off, in contradiction to the fact that R is right Noetherian. If (3.7) breaks off, at the r-th stage say, then

$$a_{r+1} = \sum_{l}^{r} a_i b_i,$$

for some elements b_i in R; it follows that $f_{r+1} - \sum f_i x^{n_{r+1} - n_r} b_i$ is a polynomial in A, of degree less than n_{r+1} but not in $\sum_1^r f_i S$. This contradicts the definition of f_{r+1}, so the chain (3.7) must break off and the result follows. ■

This result is usually stated and proved for commutative rings, but as the above proof shows, the commutativity is not used anywhere. We obtain an important case by taking R to be a field; being a simple ring, a field is obviously Noetherian, so by induction on n we obtain

Corollary 3.4

For any field k, the polynomial ring $k[x_1, \ldots, x_r]$ is a Noetherian ring. ■

The observant reader will have noticed that if in (3.5) we do not restrict the a_i to vanish for large i, we again get a ring; its elements may be written as power series in x:

$$f = a_0 + a_1 x + a_2 x^2 + \ldots, \quad (a_i \in R) \tag{3.8}$$

with the same rules for addition and multiplication as before. No doubt the reader will have met power series before in analysis, where the question of convergence has to be discussed; by contrast we here allow any coefficients. The series (3.8) will be called a **formal** power series; the ring so obtained is called the **formal power series ring** in x over R and is denoted by $R[[x]]$. The **order** of f, $\mathrm{ord}(f)$, given by (3.8), is defined as the least i such that $a_i \neq 0$. This construction is often useful, even in the study of polynomial rings. For the moment we just record a result on the units in a formal power series ring.

Proposition 3.5

Let R be any ring and $R[[x]]$ the formal power series ring in x over R. Then an element $f = \sum a_i x^i$ of $R[[x]]$ is a unit if and only if a_0 is a unit in R.

Proof

Suppose that f is a unit with inverse $g = \sum b_i x^i$; since $fg = gf = 1$, we find that $a_0 b_0 = b_0 a_0 = 1$, by equating constant terms and this shows a_0 to be a unit. Conversely, assume that a_0 is a unit in R. Then $a_0^{-1} f = 1 - h$, where h has positive order; hence we can form $(1-h)^{-1} = 1 + h + h^2 + \ldots$, for h^{n+1} has order greater than n and so only contributes terms higher than x^n. It follows that the inverse of f is given by

$$f^{-1} = (1 + h + h^2 + \ldots) a_0^{-1}. \quad ■$$

To give an example, let us take a field k and form the power series ring $A = k[[x]]$. The non-units in A are just the series of positive order; they form an ideal in A, the unique maximal ideal. A ring in which the set of all non-units is an ideal, necessarily the unique maximal ideal, is called a **local ring**. Thus our finding can be stated as

Theorem 3.6

The power series ring over a field (even skew) is a local ring. ■

The power series ring can still be enlarged by allowing a finite number of terms with negative exponents; thus instead of (3.8) we have

$$\sum_{i=-r}^{\infty} a_i x^i, \qquad \text{where } a_i \in R. \tag{3.9}$$

Note that only a finite number of terms with negative exponents may be non-zero, to ensure that the product can be defined. Such a series is called a **formal Laurent series**, after Pierre Alphonse Laurent (1813–1854), who studied such series in complex function theory. The ring of formal Laurent series over R is denoted by $R((x))$.

The construction of the polynomial ring can also be carried out using negative as well as positive powers of x; thus we consider all expressions

$$g = a_{-r}x^{-r} + a_{-r+1}x^{-r+1} + \ldots + a_0 + a_1 x + \ldots + a_n x^n, \qquad \text{where } a_i \in R, \tag{3.10}$$

with the usual rules of addition and multiplication. The expression (3.10) is called a **Laurent polynomial** and the ring $R[x, x^{-1}]$ is the **ring of Laurent polynomials**.

Exercises 3.1

1. Let k be a field. Verify that any finitely generated commutative k-algebra A is a homomorphic image of the polynomial ring $k[x_1, \ldots, x_r]$, for an appropriate r. Deduce that A is Noetherian.
2. Show that for power series over an integral domain the order function satisfies the relation $\mathrm{ord}(fg) = \mathrm{ord}(f) + \mathrm{ord}(g)$.
3. Prove that for any Noetherian ring R, the power series ring $R[[x]]$ is again Noetherian. (Hint: imitate the proof of Theorem 3.3 with the degree replaced by the order function.) Do the same for the ring of Laurent polynomials and the ring of Laurent series.
4. Show that for any field F the ring of formal Laurent series, $F((x))$, is again a field.
5. Let R be a local ring whose maximal ideal is a nilideal. Find conditions for a polynomial in $R[x]$ to be a unit.

6. A ring R may be called **matrix local** if it has a unique greatest ideal J such that R/J is Artinian (and hence, being also simple, is a full matrix ring over a skew field). Describe the units of the polynomial ring $R[x]$, where R is a matrix local ring whose greatest ideal is a nilideal.

7. Show that for any local ring L the full $n \times n$ matrix ring over L is a matrix local ring.

8. Show that if L is a local ring, then $L[x \mid x^n = 0]$ is again a local ring.

9. Let R be a k-algebra with a finite-dimensional subspace V such that (i) V generates R and (ii) for any $x, y \in V, xy - yx \in V$. Show that R is Noetherian.

10. Let A be an ideal in a polynomial ring $k[x_1, \ldots, x_n]$. Show that a generating set Γ of A can be chosen so that for every $f \in A$ with leading term u there exists $g \in \Gamma$ whose leading term divides u (such a Γ is called a **Gröbner basis**, after Wolfgang Gröbner, 1899–1980).

3.2 The Euclidean Algorithm

In the study of the divisibility of the natural numbers the division algorithm plays an important role (cf. Wallace, 1998, p. 52). We recall its form: Given any two integers a, b, if $b \neq 0$, there exist integers q, r such that

$$a = bq + r, \quad 0 \leq r < |b|. \tag{3.11}$$

Finding the numbers q and r may be interpreted as a partial division of a by b, with quotient q and remainder r. Writing $a = a_0, b = a_1$, we can form a chain $a_0, a_1, a_2, \ldots$ where a_{i+1} is the remainder in the partial division of a_{i-1} by a_i. The process, called the **Euclidean algorithm** (after Euclid of Alexandria, ca. 300 BC), terminates when we reach a zero remainder, as we will, because the values $a_2, a_3, \ldots$ form a strictly decreasing sequence of positive integers; we shall find that the last non-zero remainder is the highest common factor of a and b.

It turns out that such a Euclidean algorithm also exists for the integers in certain algebraic number fields and for polynomial rings over a field. We shall therefore describe it in general terms before applying it in particular cases. Thus we have a ring R with a function θ defined on R taking non-negative integer values; this might be the absolute value in the case of $\mathbb{Z}$ or the degree in case of the polynomial ring over a field. We shall say that R has a **division algorithm** relative to θ if

DA.1 for any $a, b \in R, b \neq 0$, there exist $q, r \in R$ such that

$$a = bq + r, \quad \theta(r) < \theta(b), \tag{3.12}$$

DA.2 for all $a, b \in R^{\times}, \theta(ab) \geq \theta(a)$. (3.13)

As a matter of fact it can be shown that whenever R has a function θ satisfying (3.12), then it also has a function satisfying (3.13), but this fact will not be needed here. We shall refer to θ satisfying (3.13) as a **norm function**, and remark that it often satisfies a stronger condition such as $\theta(ab) = \theta(a)\theta(b)$ or $\theta(ab) = \theta(a) + \theta(b)$. We note that if b in (3.12) is chosen so that $\theta(b)$ has the least possible value, then no r can exist, and it follows that b must be zero. This shows that $\theta(0)$ has the least value; usually this value may be taken to be 0. A ring is said to be **Euclidean** if it is an integral domain with a division algorithm (relative to a norm function θ). Here we shall usually assume our Euclidean domains to be commutative; of course the definition also makes sense in the non-commutative case, although then one should really speak of "right Euclidean" domains. As an example of a Euclidean domain we have fields (even skew). For in a skew field K, if $a, b \in K$ and $b \neq 0$, then $a = b(b^{-1}a)$, so (3.12) holds with $r = 0$; hence the division algorithm holds in K for the function θ defined by $\theta(a) = 1$ if $a \neq 0$, while $\theta(0) = 0$.

As an important example of Euclidean domains we have the polynomial rings over a field:

Theorem 3.7

Let K be any field (even skew), and $K[x]$ the polynomial ring in an indeterminate over K. Then $K[x]$ has the division algorithm with respect to the degree as a norm function, and hence $K[x]$ is a Euclidean domain.

The proof is given by the familiar division of polynomials. Given polynomials f, g, of degrees r and s respectively, if $r < s$, there is nothing to prove, for DA.1 then holds with $q = 0$, $r = f$. If $r \geq s$, say $f = a_0x^r + \ldots, g = b_0x^s + \ldots$, then $f - gb_0^{-1}a_0x^{r-s}$ has degree less than r and by induction on r we obtain q such that $f - gq$ has degree less than s. Hence DA.1 holds, and DA.2 follows because for any f, g we have $\deg(fg) = \deg(f) + \deg(g)$. ■

Other examples of Euclidean domains are the rings of integers in certain algebraic number fields (cf. Section 3.6).

Let R be a Euclidean domain. Given $a, b \in R$, we perform the series of partial divisions:

$$a = bq_1 + r_1, \quad b = r_1 q_2 + r_2, \quad r_1 = r_2 q_3 + r_3, \quad \dots, \tag{3.14}$$

with diminishing remainder functions: $\theta(b) > \theta(r_1) > \theta(r_2) > \dots$. Since this is a sequence of non-negative integers, it must terminate, which can only happen when $r_{k+1} = 0$ for some k. To find a neat way of writing (3.14), let us introduce the matrix

$$P(x) = \begin{pmatrix} x & 1 \\ 1 & 0 \end{pmatrix}, \quad \text{with inverse } P(x)^{-1} = \begin{pmatrix} 0 & 1 \\ 1 & -x \end{pmatrix}.$$

The equations (3.14), with the final equation $r_{k-1} = r_k q_{k+1}$ now become

$$(a, b) = (b, r_1)P(q_1), \quad (b, r_1) = (r_1, r_2)P(q_2), \quad \dots, \quad (r_{k-1}, r_k) = (r_k, 0)P(q_{k+1}).$$

They can be summed up in the single equation

$$(a, b) = (r_k, 0)P(q_{k+1})P(q_k)\dots P(q_1), \tag{3.15}$$

with its inverse

$$(r_k, 0) = (a, b)P(q_1)^{-1}\dots P(q_{k+1})^{-1}.$$

We observe that $P(x)^{-1} = JP(x)J$, where $J = \begin{pmatrix} 0 & 1 \\ -1 & 0 \end{pmatrix}$; since $J^2 = -I$, we can rewrite this equation as

$$(r_k, 0) = (-1)^k (a, b)JP(q_1)P(q_2)\dots P(q_{k+1})J. \tag{3.16}$$

Equations (3.15) and (3.16) provide an explicit relation between the elements a, b and the least non-zero remainder r_k of the Euclidean algorithm.

For the natural numbers the Euclidean algorithm is used to derive the existence of a highest common factor and a least common multiple for any two numbers. We shall now see how this can be proved quite generally. Let R be a commutative integral domain. Given $a, b \in R$, we shall write $a|b$ and say that a **divides** b if $b = ca$ for some $c \in R$. Given any elements $a, b \in R$, by a **highest common factor (HCF)** (sometimes also called "greatest common divisor") we understand an element $d \in R$ such that (i) $d|a$, $d|b$ and (ii) for any $c \in R$, $c|a$, $c|b$ implies $c|d$. Two elements a, b are said to be **coprime** if their HCF is 1. We shall denote the HCF of a and b by $\mathrm{HCF}(a, b)$ or simply (a, b) and note that it is determined up to a unit factor in R; moreover, if d is an HCF of a and b, then so is ud, for any unit u in R. Similarly, a **least common multiple (LCM)** of a and b, denoted by $\mathrm{LCM}[a, b]$ or $[a, b]$, is defined as an element $m \in R$ such that $a|m$, $b|m$ and for any $c \in R$, if $a|c$, $b|c$, then $m|c$. Again it is clear that $[a, b]$ is unique up to a unit factor. Corresponding definitions can be given for the HCF and LCM of any subset of R, but this will not be needed.

As a first application we shall prove the existence of an HCF and LCM, using (3.15) and (3.16).

Theorem 3.8

Let R be any commutative integral domain which is Euclidean. Then any two elements $a, b \in R$ have an HCF and an LCM, and they satisfy the equation

$$(a,b)[a,b]R = abR.$$

Proof

We may suppose that a, b are both non-zero. Writing $d = r_k$ for the last non-zero remainder in the Euclidean algorithm, we can express the Eqs (3.15) and (3.16) as

$$a = da', \qquad b = db', \tag{3.17}$$

$$d = ar + bs. \tag{3.18}$$

We claim that d is an HCF and ab/d is an LCM of a, b. For by (3.17), d is a common factor of a and b, and if $c|a, c|b$, then $c|d$ by (3.18). This shows d to be an HCF. Now let $m = ab/d$; then by (3.17) we have $m = ab' = ba'$, hence m is a common multiple, and if c is another common multiple, then $a|c$, $b|c$, hence $ab|bc$, $ab|ac$, and so $ab|(ar + bs)c$, i.e. $ab|dc$ or $m|c$; this shows m to be an LCM. ■

The equation (3.18), expressing the HCF of a and b as a linear combination of a and b, is called **Bezout's identity** (after Étienne Bezout, 1730–1783).

The Euclidean algorithm for R also provides information on the structure of matrix groups over R. Given any ring R, we can form the set $\mathrm{GL}_n(R)$ of all invertible $n \times n$ matrices over R; clearly this is a group under matrix multiplication, called the **general linear group** of degree n over R. For example, a diagonal matrix $\mathrm{diag}(a_1, \ldots, a_n)$ is in $\mathrm{GL}_n(R)$ iff $a_1, \ldots, a_n$ are all units in R. Among the non-diagonal elements we have the **elementary matrices**, i.e. matrices of the form

$$B_{ij}(a) = I + ae_{ij}, \qquad \text{where } a \in R,\ i \neq j.$$

It turns out that for a Euclidean domain these types of matrices already generate the general linear group:

Theorem 3.9

If R is any Euclidean domain, then the general linear group of any degree, $\mathrm{GL}_n(R)$ is generated by all elementary and diagonal matrices (with unit entries along the main diagonal).

Proof

Let $\mathrm{GE}_n(R)$ be the subgroup of $\mathrm{GL}_n(R)$ generated by all elementary and diagonal matrices. We observe that the 2×2 matrices on the right of (3.15) and (3.16) are in $\mathrm{GE}_2(R)$, because

$$P(x) = \mathrm{diag}[1, -1]B_{12}(1-x)B_{21}(-1)B_{12}(1);$$

hence for any $a, b \in R$ the row vector (a, b) can be written as $(d, 0)P$, where $P \in \mathrm{GE}_2(R)$. Given two vectors $u, v \in R^n$, let us write $u \equiv v$ and call u **congruent** to v if $u = vS$ for some $S \in \mathrm{GE}_n(R)$. Clearly this is an equivalence relation, and what we have just shown can be expressed by saying that every vector in R^2 is congruent to a vector with at most one non-zero component. This clearly also holds for vectors in R^n, for any vector $(u_1, \ldots, u_n)$ is congruent to a vector of the form $(u_1', 0, u_3', \ldots, u_n')$, by what has been shown, and by operating in turn on the first and third component, then the first and fourth component and so on, we find a vector $(u_1'', 0, \ldots, 0)$ which is congruent to the original vector.

We shall apply the result to matrices; two matrices A and B in $\mathrm{GL}_n(R)$ will be called **congruent** if $A = BQ$ for some $Q \in \mathrm{GE}_n(R)$. Now starting from any matrix $A \in \mathrm{GL}_n(R)$, we can find a matrix A' congruent to A in which the first row has all entries after the first equal to zero. Since A is invertible, so is A' and hence its (1,1)-entry must be a unit u say; if we now multiply on the right by $\mathrm{diag}(u^{-1}, 1, \ldots, 1)$, we obtain a matrix A'' congruent to A with first row $(1, 0, \ldots, 0)$. We now have

$$A'' = \begin{pmatrix} 1 & 0 & \ldots & 0 \\ * & & A_1 & \end{pmatrix},$$

where $A_1 \in \mathrm{GL}_{n-1}(R)$. By induction on n we find that $A_1 \cong I$, and now by subtracting an appropriate multiple of the i-th column from the first column, for $i = 2, \ldots, n$, we can reduce the first column to zero except for the first entry, and so find that $A \equiv I$, i.e. $A \in \mathrm{GE}_n(R)$, as we had to show. ∎

Of course the Euclidean hypothesis cannot be omitted from Theorem 3.9. For example, it can be shown that for the polynomial ring $k[x, y]$ the matrix $\begin{pmatrix} 1+xy & x^2 \\ -y^2 & 1-xy \end{pmatrix}$, which is clearly invertible, cannot be written as a product of elementary and diagonal matrices (see Cohn, 1966). This result also shows (by Theorem 3.9) that $k\,[x, y]$ is not Euclidean.

Exercises 3.2

1. Let R be a commutative integral domain with a norm function θ satisfying DA.2 and instead of DA.1 the condition A.1: Given a, b such that $b \neq 0$ and $\theta(a) \geq \theta(b)$, there exists $q \in R$ such that $\theta(a - bq) < \theta(a)$. Show that R has the division algorithm relative to θ and so is Euclidean.

2. Show that the quotient and remainder in the division of polynomials in Theorem 2.2 are uniquely determined by f and g.

3. In the Euclidean algorithm for $a, b \in \mathbb{Z}$ show that for $a \geq b \geq 1$, the k-th remainder is 0 when $k > \log a/\log 2$. (Hint: show that $r_1 < a/2$.)

4. Show that if a ring R has a Euclidean algorithm relative to a function which is constant on $R^\times$, then R is a skew field.

5. Let R be a commutative ring and for any polynomial $f \in R[x]$, denote by $f(c)$ the element of R obtained by replacing x by $c \in R$. Prove the remainder formula for the division of f by $x - c : f = (x - c)q + f(c)$, and deduce a condition for f to be divisible by $x - c$.

6. Show that for any variables $t_1, \ldots, t_n$ the product of the matrices $P(t_i)$ is given by

$$P(t_1)\ldots P(t_n) = \begin{pmatrix} p(t_1, \ldots, t_n) & p(t_1, \ldots, t_{n-1}) \\ p(t_2, \ldots, t_n) & p(t_2, \ldots, t_{n-1}) \end{pmatrix},$$

where $p(t_1, \ldots, t_n)$ is defined as the sum of $t_1 t_2 \ldots t_n$ plus all terms obtained from this one by omitting one or more pairs of adjacent factors $t_i t_{i+1}$ (this is known as the **leapfrog construction** and p is also called the **continuant polynomial**, since it occurs in the study of continued fractions).

7. Define a sequence $p_{-1}, p_0, p_1, \dots$ of polynomials in $t_1, t_2, \dots$ recursively by the rules: $p_{-1} = 0, p_0 = 1, p_n(t_1, \dots, t_n) = p_{n-1}(t_1, \dots, t_n - 1)t_n + p_{n-2}(t_1, \dots, t_{n-2})$. Verify that p_n $(t_1, \dots, t_n)$ coincides with the function $p(t_1, \dots, t_n)$ defined in Exercise 3. What is the number of terms in p_n?

8. Show that for any local ring R, $\mathrm{GE}_n(R) = \mathrm{GL}_n(R)$.

9. Show that the interchange of two columns of a matrix and changing the sign of one of them can be accomplished by right multiplication by a product of elementary matrices, but without the sign change this is not possible.

10. Let R be a Euclidean domain. Show that the ring $R((x))$ of formal Laurent series is also Euclidean. (Hint: use the order function instead of the degree.)

3.3 Factorization

No doubt all readers are familiar with the factorization of an integer into prime numbers. When the integers are introduced, it is shown that every integer $n > 1$ can either be factorized: $n = rs$, where the factors satisfy $r, s > 1$, or no such factorization is possible; in the latter case n is called a **prime number** (cf. Wallace, 1998, p. 49). Further it is shown that every positive integer has a **complete factorization**, i.e. it can be expressed as a product of primes, and any two complete factorizations differ only in the order of the prime factors. This result is known as the **fundamental theorem of arithmetic** (see Wallace, 1998, p. 180); its proof depends on the Euclidean algorithm and we shall see that a corresponding result holds for any Euclidean domain (cf. Exercise 1). The result is set as an exercise, because a more general result will be proved in the next section. The main difference between $\mathbb{Z}$ and the general case is that whereas in $\mathbb{Z}$ there are only two units, namely 1 and -1, and in $\mathbb{Z}^+$, the positive part, 1 is the only unit, a Euclidean domain may have many units.

In any commutative integral domain R we have the notion of divisibility: $a|b$ if $b = ac$ for some $c \in R$. It is easy to check that this relation is reflexive: $a|a$, and transitive: if $a|b$ and $b|c$, then $a|c$. Two elements a and b are said to be **associated** if $a|b$ and $b|a$. Associated elements can be described as follows.

Proposition 3.10

In any commutative integral domain R two elements a and b are associated if and only if $b = au$, where u is a unit in R.

Proof

If $b = au$ with a unit u, then $a \mid b$ and since $a = bu^{-1}$, we also have $b \mid a$, so a and b are associated. Conversely, if a and b are associated, then $a = bc$ and $b = ad$, hence either both a and b vanish, or neither does, so we may assume that $a, b \neq 0$. We have $a = bc = adc$, hence $a(1 - dc) = 0$; since we are in an integral domain, it follows that $dc = 1$, so c is a unit, with inverse d. ∎

To study factorization in integral domains, we shall limit ourselves to the commutative case, so unless there is a statement to the contrary (mainly in the exercises), in this chapter all integral domains will be commutative. Our first task is to define an analogue of a prime number; it turns out that in general there are two distinct notions, which are related but not identical. In an integral domain R we understand by a **prime** an element p of R, not zero or a unit, such that $p|ab$ implies $p \mid a$ or $p \mid b$. Given a prime p, suppose we have a factorization: $p = ab$. Then $p \mid ab$ and since p is a prime, we have either $p \mid a$ or $p \mid b$, say the former. Then $a = pc = abc$, hence $bc = 1$, so b is a unit and p is associated to a. We see that primes have the expected property of being "unfactorable", but this property may be shared by other elements. So we define: an element c of an integral domain R is **irreducible** or an **atom** if it is not zero or a unit and in any factorization, $c = ab$, one of the factors must be a unit. What we have seen is that every prime is an atom, but the converse may not be true; in fact we shall soon see that the converse, with another condition, just characterizes the domains which have unique factorization in the following sense: an integral domain is called a **unique factorization domain** (**UFD** for short) if every element not zero or a unit can be written as a product of atoms, and given two complete factorizations of the same element,

$$c = a_1 \dots a_r = b_1 \dots b_s, \quad \text{where } a_i, b_j \text{ are atoms,} \tag{3.19}$$

we have $r = s$ and the b's can be renumbered so that a_i is associated to b_i. The number of terms in a complete factorization of c is called the **length** of c.

The following criterion makes it easy to recognize UFDs.

Theorem 3.11

An integral domain R is a UFD if and only if

(i) every element not zero or a unit has a complete factorization;

(ii) every atom is prime.

Moreover, condition (i) holds in every Noetherian domain.

Proof

Assume that R is a UFD. Then (i) holds. To prove (ii), suppose that c is an atom such that $c|ab$, and let us take complete factorizations of a and b, $a = a_1 \dots a_r$, $b = b_1 \dots b_s$, where a_i, b_j are atoms. Then we have

$$a_1 \dots a_r b_1 \dots b_s = ab = cd,$$

for some $d \in R$. By taking a complete factorization of d, we obtain another complete factorization of ab, and by uniqueness we find that the atom c must be associated to one of the a_i or b_j, hence $c|a$ or $c|b$ and so c is a prime, as claimed.

Conversely, assume that R satisfies (i) and (ii) and take two complete factorizations of an element

$$c = a_1 \dots a_r = b_1 \dots b_s. \tag{3.20}$$

We have to show that $r = s$ and after suitably renumbering the b's, b_i is associated to a_i. We shall use induction on r. When $r = 1$, then s must also be 1 and there is nothing to prove, so let $r > 1$. By hypothesis, a_1 is an atom, hence prime and $a_1 | b_1 \dots b_s$, so $a_1 | b_j$ for some j, say $j = 1$, by renumbering the b's. Now b_1 is also an atom, hence a_1 and b_1 are associated, say $a_1 = b_1 u$, where u is a unit. Dividing (3.20) by b_1 we get

$$ua_2 \dots a_r = b_2 \dots b_s.$$

By induction, $r - 1 = s - 1$ and we can renumber the b's so that a_i is associated to b_i for $i = 2, \dots, r$. This has also been shown to hold for $i = 1$, and it proves that R is a UFD.

Finally to prove the last part, assume that R is a Noetherian domain. Given $c \in R$, where c is not zero or a unit, either c is an atom or it has a factorization $c = a_1 b_1$ into non-units. If b_1 is not an atom, we can continue the factorization: $b_1 = a_2 b_2$, hence $c = a_1 a_2 b_2$ and continuing in this way, we obtain a strictly ascending chain of ideals:

$$(c) \subset (b_1) \subset (b_2) \subset \dots;$$

since R is Noetherian, this chain must break off, which is possible only if b_n is a unit and a_n an atom. This shows that any non-unit has an atomic factor. We now carry out the same process, but choosing an atom a_i at each stage; again the process must end and when it does, we have a complete factorization of c. ■

In any UFD R, any two elements a, b have an HCF and an LCM, which can be determined as follows: we take complete factorizations of a and b:

$$a = a_1 \dots a_r, \qquad b = b_1 \dots b_s, \qquad a_i, b_j \text{ atoms.} \tag{3.21}$$

Let us number the a's and b's so that any associated pairs stand at the beginning; thus we may assume that a_i is associated to b_i for $i = 1, \dots, k$, but that no a_i for $i > k$ is associated to any b_j for $j > k$. Then $d = a_1 \dots a_k$ is an HCF of a and b; for it is clearly a factor of a and also of b, and if c is a factor of a and b, then any atomic factor p of c must divide a and b and so will be associated to a_i for some $i \leq k$, say a_1. Put $c = pc'$; then $c' \,|\, a_2 \dots a_r$, $c' \,|\, b_2 \dots b_s$, hence by induction on k, we have $c' \,|\, a_2 \dots a_k$ and so $c | d$, which shows d to be an HCF of a and b. In the same way it can be shown that a and b have an LCM, and in fact if d is an HCF and m an LCM, then $dm = abu$ for some unit u. The result may be stated as

Proposition 3.12

Let R be a UFD. Then any two elements a, b have an HCF d and an LCM m and dm is associated to ab.

To complete the proof we need only check what happens if one or both of a, b are zero. Suppose that $b = 0$; the HCF of a and 0 is a and the LCM is 0, and $a.0 = 0$, so the result continues to hold in this case. ∎

One of the main tools in finding UFDs is the result that the polynomial ring over a UFD is again a UFD. This will be proved in the next section; for the moment we shall bring a result that will be needed. First a definition: if R is a UFD and $R[x]$ the corresponding polynomial ring, then a polynomial $f \in R[x]$ is said to be **primitive** if its coefficients have no common factor apart from units. The next result is known as **Gauss's lemma**:

Theorem 3.13

The product of two primitive polynomials over a unique factorization domain is again primitive.

Proof

Let R be a UFD and $f = \sum a_i x^i$, $g = \sum b_j x^j$ be primitive polynomials over R and suppose that

$$fg = dh, \tag{3.22}$$

where d is a non-unit in R and $h \in R[x]$. Take a prime p dividing d; since f is primitive, not all its coefficients a_i are divisible by p, say a_r is the first coefficient of f not divisible by p. Similarly, let b_s be the first coefficient of g not divisible by p. If we put $h = \sum c_k x^k$ and compare the coefficients of x^{r+s} in (3.22), we find

$$dc_{r+s} = a_0 b_{r+s} + a_1 b_{r+s-1} + \ldots + a_r b_s + \ldots + a_{r+s} b_0.$$

All terms on the right, except possibly $a_r b_s$ are divisible by p, as is dc_{r+s}, hence $p \,|\, a_r b_s$, and so either $p \,|\, a_r$ or $p \,|\, b_s$, but this contradicts the hypothesis and so the conclusion follows. ■

Exercises 3.3

1. Using the fact that the HCF of a and b can be written in the form $ar + bs$, show that in a Euclidean domain every atom is prime; deduce that every Euclidean domain is a UFD.

2. In any commutative integral domain show that if a, b have the LCM m, then ca, cb have the LCM cm.

3. In a UFD, if d is an HCF of a and b, show that ab/d is an LCM of a and b (i.e. carry out the part of the proof of Proposition 3.12 that was omitted).

4. Let A be a commutative integral domain. Show that if two elements $a, b \in A$ have an LCM, then they have an HCF (see also Exercise 3.4.6 below).

5. Define HCF and LCM for any finite subsets of an integral domain. Can the definition be extended to infinite subsets?

6. Let k be any field and consider the polynomial ring in infinitely many indeterminates $R = k[x_1, x_2, \ldots]$. Show that every element of R has a complete factorization, even though R is not Noetherian (below in Section 3.4 we shall see that R is a UFD).

7. Show that the power series ring $R[[x]]$ over a UFD R satisfies Gauss's lemma (Theorem 3.13) and deduce that $R[[x]]$ is a UFD.

8. Let R be a commutative Noetherian integral domain which is not a UFD. Given two inequivalent factorizations of an element c of R, find an atom which is not prime.

9. In a non-commutative integral domain R define b to be a left factor of a if $a = bc$ for some $c \in R$, and call a and b **right associated** if each is a left factor of the other. Show that a, b are right associated iff $a = bu$ for some unit u in R.

10. Let R be an integral domain (not necessarily commutative). Define atoms in R and show that if R is left and right Noetherian, then every non-zero non-unit can be written as a finite product of atoms.

3.4 Principal Ideal Domains

In Section 3.2 we met the class of Euclidean domains, and found that the existence of a division algorithm has a number of useful consequences. Nevertheless, the definition is somewhat artificial, and it will be convenient to consider the slightly larger class of principal ideal domains. This class, soon to be defined, includes all Euclidean domains and it shares many important properties with this subclass. Its advantage is that it has an intrinsic definition which does not involve a norm function.

In any ring R a right ideal is said to be **principal** if it can be generated by a single element; thus it has the form cR, where $c \in R$. Similarly a left ideal of the form Rc is called **principal**, but here we shall mainly be concerned with the commutative case. We now define a **principal ideal domain (PID)** as a commutative integral domain in which every ideal is principal. By definition every ideal in such a ring is finitely generated, so by Theorem 2.2 every PID is Noetherian. Our first task is to show that Euclidean domains are principal:

Theorem 3.14

Every Euclidean domain is a principal ideal domain.

Proof

Let R be a Euclidean domain with norm function θ and I any ideal in R. When $I = 0$, there is nothing to prove, so let $I \neq 0$ and choose $c \in I$ such that $c \neq 0$ and subject to these conditions, $\theta(c)$ has its least value. We have $(c) \subseteq I$ and we claim that equality holds here. For if $b \in I$, then by the division algorithm we have $b = cq + r$, where $\theta(r) < \theta(c)$, but $r = b - cq \in I$ and since $\theta(c)$ was minimal, we must have $r = 0$. It follows that $b = cq$, and this shows that $(c) = I$ as claimed; thus I is principal and it follows that R is a PID. ■

We next show that every PID is a UFD, using the criterion of Theorem 3.11.

Theorem 3.15

Every principal ideal domain is a unique factorization domain.

Proof

Let R be a PID; to show that it is a UFD, it is enough to verify conditions (i) and (ii) of Theorem 3.11 and here (i) follows because R is Noetherian. To verify (ii) let p be an atom and suppose that $p \mid ab$. The ideal generated by p and a is principal, say $(p, a) = (d)$. Since p is an atom, d as a factor of p is either associated to p, in which case it may be taken to be equal to p, or it is a unit, and so may be taken to be 1. If $d = p$, then $p \mid a$, while if $d = 1$, then $1 = ar + ps$, hence $b = abr + pbs$. Now $p \mid ab$, say $ab = pt$; it follows that $b = ptr + pbs = p(tr + bs)$, and so $p \mid b$, so p is indeed a prime. We can now apply Theorem 3.11 and conclude that R is a UFD. ■

Bearing in mind Theorem 3.14, we obtain the following consequence:

Corollary 3.16

Every Euclidean domain is a unique factorization domain. In particular, this always holds for the polynomial ring over a field. ■

We can now also prove that the polynomial ring over a UFD is again a UFD, even though it may not be a PID.

Theorem 3.17

If R is a unique factorization domain, then so is the polynomial ring $R[x]$. More generally, $R[x_1, \ldots, x_n]$ is a unique factorization domain for any $n \in \mathbb{N}$.

Proof

Let K be the field of fractions of R. By Corollary 3.16, $K[x]$ is a UFD; we shall show that $R[x]$ is a UFD by verifying the conditions of Theorem 3.11. Let $f = \sum a_i x^i$ be any element of $R[x]$; a complete factorization of f over $K[x]$ can be written as

$$df = q_1 \ldots q_r, \tag{3.23}$$

where $d \in R$, $q_i \in R[x]$, by clearing the denominators. If d is not a unit, take a prime factor p of d; this divides $q_1 \dots q_r$, and by Gauss's lemma (Theorem 3.13), p must divide one of the q's, say $p \,|\, q_1$. But then we can replace q_1 by q_1/p and so omit p from (3.23). After a finite number of steps we have reduced d to a unit in R and now we can take it to the other side. To obtain a complete factorization over R, we still have to split off any prime factors in R from each q_i, but since such a factor will have to divide each coefficient of q_i, there can only be finitely many and we thus obtain a complete factorization of f over $R[x]$.

Secondly we have to show that each atom in $R[x]$ is a prime. For the atoms of R this follows by Gauss's lemma, so let $h = \sum c_i x^i$ be an atom of positive degree in x and assume that $h \,|\, fg$ in $R[x]$, i.e. $fg = hk$, where $k \in R[x]$. Since $K[x]$ is a UFD, we have $h \,|\, f$ or $h \,|\, g$ in $K[x]$, say the former. Clearing denominators, we have

$$df = hq, \quad \text{where } d \in R^{\times}, \quad q \in R[x]. \tag{3.24}$$

If d is not a unit, take a prime divisor p of d. Then $p \,|\, hq$, hence $p \,|\, h$ or $p \,|\, q$ in $R[x]$, by Gauss's lemma. Now h is an atom and is not associated to an element of R, so we must have $p \,|\, q$; but then we can replace q by q/p and continue the reduction. After a finite number of steps d is reduced to a unit in R and so can be brought to the other side of (3.24), and this shows that $h \,|\, f$, as claimed. Now the final part follows by induction on n. ■

In Theorem 3.9 we obtained an expression for invertible matrices over a Euclidean domain, and it would be useful to have something similar for PIDs, but examples show that the corresponding fact does not hold for all PIDs (cf. Cohn, 1966). However, there is a useful normal form for arbitrary matrices over a PID, which we shall now derive.

Theorem 3.18

Let R be a principal ideal domain, and consider an $m \times n$ matrix A over R. Then there exist invertible matrices P, Q of orders m, n respectively over R such that

$$PAQ = \begin{pmatrix} d_1 & 0 & 0 & & \dots & 0 \\ 0 & d_2 & 0 & & \dots & 0 \\ 0 & 0 & \dots & d_r & \dots & 0 \\ 0 & 0 & \dots & & \dots & 0 \end{pmatrix} \tag{3.25}$$

where $d_i \,|\, d_{i+1}$ for $i = 1, \dots, r-1$; more briefly, $PAQ = \mathrm{diag}(d_1, d_2, \dots, d_r, 0, \dots, 0)$.

Proof

The proof resembles that of Theorem 3.9; we begin with a reduction of 2-component vectors, but we shall now call two vectors $\boldsymbol{u}$, $\boldsymbol{v}$ **right associated** if $\boldsymbol{u} = \boldsymbol{v}S$ for some $S \in \mathrm{GL}_2(R)$.

We claim that any vector (a, b) is right associated to $(h,0)$, where h is an HCF of a and b. Since R is a PID, a and b have an HCF h, say $a = ha'$, $b = hb'$, and since h generates the ideal generated by a and b, we have $h = ad' - bc'$, or dividing by h, we find $1 = a'd' - b'c'$. Hence

$$(a \quad b) = (h \quad 0)\begin{pmatrix} a' & b' \\ c' & d' \end{pmatrix}, \qquad (h \quad 0) = (a \quad b)\begin{pmatrix} d' & -b' \\ -c' & a' \end{pmatrix}.$$

and this shows $(a,\ b)$ to be right associated to $(h,\ 0)$.

Turning now to the general case, we can as in the proof of Theorem 3.9, find a matrix right associated to A in which all entries of the first row after the first one are zero, and continuing in this way, we find a matrix $B = (b_{ij})$ right associated to A, such that $b_{ij} = 0$ for $i < j$. We shall now have to operate with invertible matrices on the left, and this suggests calling two matrices A and A' **associated** if $A' = SAT$, for invertible matrices S and T of appropriate size. By multiplying B on the left by an invertible matrix we can reduce the entries of the first column after the first one to zero. This may introduce more non-zero elements in the first row, but if it does, it will reduce the length of the (1,1)-entry, so by an induction on the length of a_{11} we find that A is associated to

$$\begin{pmatrix} a_1 & 0 & \dots & 0 \\ 0 & & & \\ \cdot & & A_1 & \\ \cdot & & & \end{pmatrix}$$

where A_1 is an $(m-1) \times (n-1)$ matrix. An induction on $m+n$ shows that A_1 is associated to a matrix in diagonal form, say $\mathrm{diag}(a_2, a_3, \dots, a_r, 0, \dots, 0)$. By combining this with the previous statement we find that A is associated to a diagonal matrix:

$$SAT = \mathrm{diag}(a_1, a_2, \dots, a_r, 0, \dots, 0). \tag{3.26}$$

If $a_i \mid a_{i+1}$ for $i = 1, \dots, r-1$, this is of the required form, so assume that a_1 does not divide a_2 and consider the 2×2 matrix formed by the first two rows and columns. We have the equation

$$\begin{pmatrix} 1 & 1 \\ 0 & 1 \end{pmatrix}\begin{pmatrix} a_1 & 0 \\ 0 & a_2 \end{pmatrix} = \begin{pmatrix} a_1 & a_2 \\ 0 & a_2 \end{pmatrix};$$

if we reduce the matrix on the right to diagonal form as before we again reach the form (3.26) but with a_1 of shorter length than before. After a finite number of steps we have $a_1 \mid a_2$ and by repeating this procedure we ensure that $a_1 \mid a_i$ for $i = 2, \ldots, r$. The same process can be carried out for $a_2, \ldots, a_{r-1}$, and so we finally reach the form (3.23). ■

This reduction to diagonal form is called the **PAQ-reduction**, or also the **Smith normal form**, after Henry John Stephen Smith (1826–1883), who first carried out this reduction over the ring of integers in 1861. The extension to PIDs is more recent, but the most important case remains that of Euclidean domains. A similar reduction can be carried out for non-commutative PIDs, but then the notion of divisibility turns out to be more complicated (cf. A.3, Section 9.2).

There remain the questions: (i) to what extent is the form (3.25) unique; and (ii) how can the d_i in (3.25) can be determined in terms of A (without carrying out the full reduction)? Both questions may be answered as follows. We recall that for any matrix A a **k-th order minor** is the determinant of the submatrix obtained by taking k rows and k columns of A. Over a PID we can form Δ_k, the HCF of all k-th order minors of A; it is called the k-th **determinant divisor** of A. Any determinant divisor of a matrix product AB is a multiple of the corresponding determinant divisor of A, as the expansion of the product shows. It follows that the determinant divisors of any matrix associated to A are associated to those of A, and a glance at the form (3.23) shows that the k-th determinant divisor Δ_k of A is given by

$$\Delta_k = d_1 d_2 \ldots d_k, \quad k = 1, \ldots, r. \tag{3.27}$$

Hence the elements d_i in (3.23) are given by the equations

$$d_i = \Delta_i / \Delta_{i-1}, \quad i = 1, \ldots, r. \tag{3.28}$$

Like the Δ_i, the d_i are unique up to associates; they are known as the **invariant factors** of the matrix A.

Exercises 3.4

1. Given a PID R and $a, b \in R$, show that the HCF d and LCM m of a and b are given by $aR + bR = dR$, $aR \cap bR = mR$. By applying the Second Isomorphism Theorem, deduce the relation $abR = dmR$.

2. Let R be a UFD which is a k-algebra (where k is a field), such that every unit in R belongs to k, but $R \neq k$. Show that R contains infinitely many atoms that are not associated.

3. Let R be a PID and K its field of fractions. Show that every element u of K can be written as $u = a/b$, where $b \neq 0$ and a, b are coprime; to what extent are a, b determined by u?

4. Let R be a PID and T a multiplicatively closed subset of $R^{\times}$. Show that the localization R_T is again a PID.

5. Let R be a commutative integral domain such that (i) every finitely generated ideal is principal and (ii) R satisfies the maximum condition for principal ideals. Show that R is a PID.

6. Let A be the subring of $\mathbb{Z}[x]$ consisting of all polynomials with even coefficient of x. Show that A is not a PID and find two elements without an LCM.

7. Find a PAQ-reduction for $\begin{pmatrix} 1 & 4 & 7 & 9 \\ 3 & 14 & 27 & 45 \\ 2 & 8 & 24 & 48 \end{pmatrix}$ and determine its determinant divisors.

8. Let A be a matrix over $\mathbb{Z}$ with determinant divisors $\Delta_1, \ldots, \Delta_r$. Find a condition on the invariant factors for a determinant divisor Δ_i to exist containing no prime to the first power only.

9. Let F be an infinite field and consider a matrix $A = A(x)$ over the polynomial ring $F[x]$. Show, by performing a PAQ-reduction, that the rank of A over the field of fractions $F(x)$ is the maximum of the ranks of $A(c)$ for all $c \in F$. (Hint: use the fact that a polynomial over F has only finitely many zeros.)

10. If A, B are square matrices of the same size over a PID, such that $AB = I$, show that $BA = I$.

3.5 Modules over Principal Ideal Domains

We have seen that abelian groups may be regarded as $\mathbb{Z}$-modules; thus modules form a generalization of abelian groups, and our aim in this section is to extend theorems on abelian groups to modules over a PID. In particular we shall find a generalization of the fundamental theorem of abelian groups, stating that every finitely generated abelian group is a direct sum of cyclic groups.

From Section 1.4 we recall that any module over a general ring R is a homomorphic image of a free module, and that every free R-module can be expressed as a direct sum of copies of R. Using the language of exact sequences, we can say that every R-module M forms part of a short exact sequence

$$0 \to G \to F \to M \to 0, \tag{3.29}$$

where F is free. This is called a **presentation** of M; if M is finitely generated, F can be taken to be of finite rank. If G is finitely generated, M is said to be **finitely related**, while M is said to be **finitely presented** by (3.29) if both F and G are finitely generated.

If R is any (commutative) integral domain and M a right R-module, then an element $u \in M$ is called a **torsion element** if $uc = 0$ for some $c \in R^{\times}$; the set of all torsion elements of M is a submodule, for if $uc = 0$, then $(ux)c = ucx = 0$ for any $x \in R$ and if $uc = vd = 0$, then $(u - v)cd = 0$. This submodule is called the **torsion submodule** of M and is denoted by $\mathrm{t}M$; further, M is called a **torsion module** if $\mathrm{t}M = M$ and **torsion-free** if $\mathrm{t}M = 0$. In order to obtain a decomposition of modules over a PID, we first find a criterion for free modules. It will be convenient to use left modules in the proof, though of course the result is left–right symmetric.

Lemma 3.19

Let R be a principal ideal domain. Then a finitely generated R-module is free if and only if it is torsion-free.

Proof

It is clear that every free R-module is torsion-free; in fact this holds more generally over any integral domain. Conversely, let M be a torsion-free left module over a PID R, which is finitely generated, by $u_1, \ldots, u_n$ say. We shall use induction on n; for $n = 1, M = Ru_1$ and the result is clear, so let $n > 1$. If M is not free on $u_1, \ldots, u_n$, then there is a non-trivial relation

$$c_1u_1 + \ldots + c_nu_n = 0. \tag{3.30}$$

Now by Theorem 3.18, with $m = 1$, there exists an invertible $n \times n$ matrix P such that

$$(c_1, \ldots, c_n)P = (d, 0, \ldots, 0). \tag{3.31}$$

Strictly speaking there will also be a 1×1 matrix on the left, but since this is a unit, it can be omitted. If we write $P = (p_{ij})$ and define v by $u_i = \sum p_{ij}v_j$, then $v_1, \ldots, v_n$ again form a generating set of M and the relation (3.30) becomes $dv_1 = 0$. Since (3.30) was non-trivial, $d \neq 0$ and hence $v_1 = 0$ (because M is torsion-free); this means that M is generated by $v_2, \ldots, v_n$ and so by induction on n, M is free and the result follows. ∎

The proof for right modules is of course the same, except that the rows in (3.31) appear as columns. In fact, the result holds even for non-commutative PIDs, though one then has to take a little more care in proving Theorem 3.18 (cf. A.3, Section 9.2).

Theorem 3.20

Let R be any integral domain and M any R-module. Then we have the following short exact sequence, where $\mathrm{t}M$, the torsion submodule of M, is a torsion module and the quotient $M/\mathrm{t}M$ is torsion-free.

$$0 \to \mathrm{t}M \to M \to M/\mathrm{t}M \to 0. \tag{3.32}$$

If, moreover, R is a principal ideal domain and M is finitely generated, then $M/\mathrm{t}M$ is free and this sequence is split exact.

Proof

The sequence (3.32) arises because the natural homomorphism $M \to M/\mathrm{t}M$ has the kernel $\mathrm{t}M$. Clearly $\mathrm{t}M$ is a torsion module; to show that $M/\mathrm{t}M$ is torsion-free, assume that α is a torsion element of $M/\mathrm{t}M$, say $c\alpha = 0$, where $c \neq 0$ in R. Take $u \in M$ mapping to α; then $cu \in \mathrm{t}M$; hence $dcu = 0$ for some $d \neq 0$. It follows that u is a torsion element and this means that $\alpha = 0$; this shows $M/\mathrm{t}M$ to be torsion-free. Now when M is finitely generated and R is a PID, then $M/\mathrm{t}M$ is free, by Lemma 3.19; if $u_1, \ldots, u_n$ is a basis of $M/\mathrm{t}M$, and v_i is an inverse image of u_i in M, then the submodule of M generated by the v_i, F say, is free, hence the natural mapping $M \to M/\mathrm{t}M$ when restricted to F is an isomorphism, therefore $M = F \oplus \mathrm{t}M$ and (3.32) is split exact. ■

In the case of a PID we shall find that every finitely generated module has a finite presentation. This will follows from our next result.

Theorem 3.21

If R is a principal ideal domain, then for any integer n, any submodule of R^n is free of rank at most n.

Proof

We shall use induction on n. For $n = 0$ the result is clear, since $R^0 = 0$* and this is free on the empty set. Now assume $n > 0$ and let F be a submodule of

*It should be clear from the context that R^0 here means the free module of rank zero.

R^n. We have the projections $\pi_i : R^n \to R$, given by $(x_1, \ldots, x_n) \mapsto x_i$. Let π'_i be its restriction to F; this is a homomorphism from F to R. Consider π'_1: either $\pi'_1 = 0$; then $F = \ker \pi'_1 \subseteq R^{n-1}$ and so the result follows by induction on n, or the image of π'_1 is a non-zero ideal of R, say im $\pi'_1 = (c) \neq 0$. Take $u \in F$ to be an inverse image of c under π'_1 and write $K = \ker \pi'_1$. We claim that

$$F = Ru \oplus K. \tag{3.33}$$

Given $x \in F$, we have $(x)\pi'_1 = yc$; hence $(x - yu)\pi'_1 = yc - yc = 0$, so $x - yu \in K$, and (3.33) follows.

Now the mapping $\varphi : x \mapsto xu$ is a homomorphism from R to Ru, which is injective, for if $zu = 0$ for some $z \in R$, then $zc = z(u)\pi'_1 = (zu)\pi'_1 = 0$, and so $z = 0$ because R is an integral domain, hence φ is an isomorphism and so Ru is free of rank 1. Now K, as submodule of R^{n-1}, is free of rank at most $n-1$, by induction, and it follows by (3.33) that F is free of rank at most $1 + (n-1) = n$, as claimed. ■

As a consequence of this theorem we may assume the module G in (3.29) to be free, at least when M is finitely generated. As a matter of fact this result holds for any R-module M. More precisely, when M is generated by n elements, then it has a presentation

$$0 \to R^m \to R^n \to M \to 0, \tag{3.34}$$

where $m \leq n$.

As a further consequence of Theorem 3.21 we have

Corollary 3.22

Given a principal ideal domain R and a finitely generated R-module M, with an n-element generating set, then any submodule of M is again finitely generated, by a set of at most n elements.

Proof

By Theorem 3.21 we have $M = F/G$, where F is free of rank n. Any submodule M' of M can be written as H/G, where H is a submodule of F containing G, as a consequence of the third isomorphism theorem. Now H is free of rank at most n, by Theorem 3.21, hence M', as a homomorphic image of H, can also be generated by n elements. ■

Finally we have the generalization of the fundamental theorem for abelian groups. This can take several forms; we shall prove one form and leave others to the exercises.

Theorem 3.23

Let R be a principal ideal domain and M a finitely generated R-module. Then M is a direct sum of cyclic modules:

$$M = R/Rd_1 \oplus \ldots \oplus R/Rd_m \oplus R^{n-m}, \qquad \text{where } d_i \,|\, d_{i+1}, \quad i = 1, \ldots, m-1. \tag{3.35}$$

Proof

Suppose that M has the presentation (3.34), where M is the cokernel of a homomorphism $\varphi : R^m \to R^n$, which is given by an $m \times n$ matrix A. By Theorem 3.18 there exist invertible matrices P, Q such that $PAQ = \text{diag}(d_1, \ldots, d_r, 0, \ldots, 0)$, where $d_i \,|\, d_{i+1}$ for $i = 1, \ldots, r-1$, and $r = m$, because φ is injective. Here P and Q correspond to changes of bases in R^m and R^n respectively; these changes do not affect the cokernel, so if $v_1, \ldots, v_n$ is the new basis in R^n, then the submodule R^m has the basis $d_1v_1, \ldots, d_mv_m$ and the cokernel takes the form on the right of (3.35). ■

Exercises 3.5

1. Let R be a PID and M a finitely generated torsion module over R. Show that M can be written as a direct sum of submodules M_p, where p runs over the different (i.e. non-associated) primes in R and M_p consists of elements that are annihilated by a power of p (a module of the form M_p is called **p-primary**).

2. Let R be a PID. Show that every finitely generated R-module can be written as a direct sum of infinite cyclic modules and cyclic p-primary modules, for varying p.

3. Let R be a PID and K its field of fractions. Show that an R-submodule of K is isomorphic to R iff it is finitely generated.

4. Let k be a field and $R = k[x, y]$ the polynomial ring in x and y over k. Find a finitely generated R-module which is torsion-free but not free.

5. Let R be a PID. Show that every matrix A over R can be written $A = TP$, where P has determinant 1 and T is lower triangular (this is known as the **Hermite reduction**, after Charles Hermite, 1822–1901). To what extent are the diagonal elements of T determined by A?

6. Let R be a PID with field of fractions K. If M is a finitely generated R-module, show that for any $x \in M$, $x \neq 0$, there exists a homomorphism $f : M \to K/R$ such that $xf \neq 0$.

7. Let R be a PID (not necessarily commutative). If $a \in R$ is represented by R/Ra, show that a left factor of a represents a submodule and a right factor a quotient of R/Ra. Hence find the conditions on $a, b \in R^\times$ to ensure that $\mathrm{Hom}_R(R/Ra, R/Rb) \neq 0$.

8. Let R be a PID. Show that an $m \times n$ matrix A has a right inverse iff $m \leq n$ and A has no non-unit square left factor.

3.6 Algebraic Integers

An important example of a principal ideal domain goes back to C.F. Gauss; the ring of **Gaussian integers** is defined as the set $G = \mathbb{Z}[i]$ of all complex numbers of the form $a + b\mathrm{i}$, where $a, b \in \mathbb{Z}$ and $\mathrm{i} = \sqrt{-1}$. Clearly it is a ring, since $(a + b\mathrm{i})(a' + b'\mathrm{i}) = (aa' - bb') + (ab' + ba')\mathrm{i}$. If we represent complex numbers in the coordinate plane, taking real numbers along the x-axis and multiples of i along the y-axis, then the elements of G are represented by all the points whose two coordinates are integers; they are said to form a **lattice** in the plane (not to be confused with an abstract lattice, which will not concern us in this volume, cf. e.g. A.2, Chapter 2).

To prove that G is a PID, we shall show that it satisfies a form of the division algorithm, relative to the norm $\mathrm{N}(a + b\mathrm{i}) = a^2 + b^2$. Given $\alpha, \beta \in G$, $\beta \neq 0$, we have to find $\lambda \in G$ such that $\mathrm{N}(\alpha - \lambda\beta) < \mathrm{N}(\beta)$. Writing $\xi = \alpha/\beta$ and dividing this condition by $\mathrm{N}(\beta)$, we can express it as follows:

$$\text{Given } \xi \in \mathbb{Q}(\mathrm{i}), \text{ there exists } \lambda \in G \text{ such that } \mathrm{N}(\xi - \lambda) < 1. \tag{3.36}$$

Here ξ is any point in the plane, say $\xi = x + y\mathrm{i}$. Let the nearest integers to x, y be a, b respectively; then $|x - a| \leq \frac{1}{2}, |y - b| \leq \frac{1}{2}$; so if $\lambda = a + b\mathrm{i}$, we have

$$\mathrm{N}(\xi - \lambda) = (x - a)^2 + (y - b)^2 \leq \tfrac{1}{4} + \tfrac{1}{4} = \tfrac{1}{2}. \tag{3.37}$$

Thus (3.36) is satisfied; with the help of this division algorithm we easily see that G is a PID, and by Corollary 3.16 it follows that G is a UFD. This method immediately suggests that we try the same for other rings of integers in $\mathbb{C}$. We observe that G was obtained by adjoining a root of the equation $x^2 + 1 = 0$ to $\mathbb{Z}$, in other words, we are dealing with a **quadratic extension** of

$\mathbb{Z}$. The general quadratic extension of $\mathbb{Z}$ is formed by adjoining to $\mathbb{Z}$ a root of a quadratic equation

$$x^2 + ax + b = 0, \quad \text{where } a, b \in \mathbb{Z}. \tag{3.38}$$

If a is even, say $a = 2a'$, then we can complete the square in (3.38) and obtain

$$(x + a')^2 - (a'^2 - b) = 0.$$

Thus our extension ring is obtained by adjoining $\sqrt{d}$ to $\mathbb{Z}$, where $d = a'^2 - b$, but there remains the question of which numbers are to be included when a is odd. This problem is overcome most easily by forming first a quadratic extension field F of $\mathbb{Q}$ and then taking all the integers in F, where $\xi \in F$ is called an **integer**, more precisely, an **algebraic integer** if it satisfies an equation with rational integer coefficients and highest coefficient 1. It is clear that every quadratic extension of $\mathbb{Q}$ is obtained by adjoining $\sqrt{d}$, for some $d \in \mathbb{Z}$; here we may assume that $d \neq 1$ and d is **square-free**, i.e. in the factorization of d at least one prime occurs and any prime occurs at most once. Now it is easy to describe the algebraic integers in $\mathbb{Q}(\sqrt{d})$:

Theorem 3.24

Any integral quadratic extension A of $\mathbb{Z}$ is obtained as the ring of algebraic integers in $\mathbb{Q}(\sqrt{d})$, where d is a square-free integer different from 1. The elements of A are all the numbers of the form $a + b\sqrt{d}(a, b \in \mathbb{Z})$ if $d \equiv 2$ or 3 (mod 4), and all numbers $a + b(1 + \sqrt{d})/2$ if $d \equiv 1 \pmod 4$.

Proof

If $\lambda = a + b\sqrt{d}$, then λ is a root of the equation

$$x^2 - 2ax + (a^2 - b^2 d) = 0. \tag{3.39}$$

Thus a is either an integer or half an odd integer, and it follows that b is either an integer or half an odd integer. Suppose first that $2a$ is odd, say $2a = 2m + 1$; then $(2a)^2 = 4m^2 + 4m + 1 \equiv 1 \pmod 4$, but $a^2 - b^2 d$ is an integer, hence $4(a^2 - b^2 d) \equiv 0 \pmod 4$, and so $4b^2 d \equiv 1 \pmod 4$. If b is an integer, then $4b^2 d \equiv 0 \pmod 4$, a contradiction with the last congruence, so $b = n + \frac{1}{2}$, $4b^2 = 4(n^2 + n) + 1$, hence $d \equiv 1 \pmod 4$. Conversely, when $d \equiv 1 \pmod 4$, all numbers of the form $a + b(1 + \sqrt{d})/2$, where $a, b \in \mathbb{Z}$, are algebraic integers. The argument also shows that when $d \equiv 2$ or 3 (mod 4), then $2a$ must be even, hence a is integral, and so b must also be an integer. Conversely, when $2a$ is even, a and hence b is integral. Of course d cannot be divisible by 4, since it is square-free. ∎

To find the quadratic fields with a Euclidean algorithm, also known as **Euclidean fields**, let us first assume that $d \equiv 2$ or $3 \pmod 4$. Any integer in the field has the form $\lambda = a + b\sqrt{d}$, with norm $\mathrm{N}(\lambda) = a^2 - b^2 d$. The condition (3.37) is now replaced by

$$|\mathrm{N}(\xi - \lambda)| = |(x-a)^2 - (y-b)^2 d| < 1. \tag{3.40}$$

Here $(x-a)^2, (y-b)^2 \leq \frac{1}{4}$, hence $|d-1| < 4$, and we obtain as the possible values (besides $d = -1$), $d = 2, 3, -2$. When $d \equiv 1 \pmod 4$, say $d = 4n+1$, (3.40) is replaced by

$$|(x - a - b/2)^2 - (y - b/2)^2 d| < 1. \tag{3.41}$$

By taking $b \in \mathbb{Z}$ closest to $2y$, we ensure that $|2y - b| \leq \frac{1}{2}$, and then choosing $a \in \mathbb{Z}$ such that $|x - a - b/2| \leq \frac{1}{2}$, we find that (3.42) holds, provided that $|\frac{1}{4} - d/16| < 1$, i.e. $|4 - d| < 16$. This holds for $d = 5, -3, -7, -11$, so we have proved

Theorem 3.25

The ring of integers in $\mathbb{Q}(\sqrt{d})$ is Euclidean for $d = 2, 3, 5, -1, -2, -3, -7, -11$. ■

By a more refined argument it can be shown that the Euclidean algorithm holds altogether for 16 positive values (cf. Stewart and Tall, 1987). As a consequence these rings of integers are unique factorization domains. The ring of integers in $\mathbb{Q}(\sqrt{d})$ is known to be a UFD for $d = -19, -43, -67, -163$ but no other negative values, and for 35 other positive values less than 100, but it is not even known whether it holds for infinitely many values of n.

We conclude with an example where unique factorization fails to hold. Consider the ring A of integers in $\mathbb{Q}(\sqrt{-5})$. Since $-5 \equiv 3 \pmod 4$, we see from Theorem 3.24 that the elements of A are $a + b\sqrt{-5}$, where $a, b \in \mathbb{Z}$. In A the number 6 has the factorizations

$$6 = 2.3 = (1 + \sqrt{-5})(1 - \sqrt{-5}); \tag{3.42}$$

further, $\mathrm{N}(2) = 4$, $\mathrm{N}(3) = 9$, $\mathrm{N}(1 \pm \sqrt{-5}) = 6$. If, for example, $1 + \sqrt{-5}$ could be factorized as $\alpha\beta$, then $\mathrm{N}(\alpha)$, $\mathrm{N}(\beta)$ must be 2 and 3, say $\mathrm{N}(\alpha) = 2$, $\mathrm{N}(\beta) = 3$, but the equation $x^2 + 5y^2 = 2$ or 3 has no integer solutions, as is easily seen. Hence $1 + \sqrt{-5}$ is an atom, and the same argument applies to 2, 3 and $1 - \sqrt{-5}$. Thus (3.42) represents two essentially distinct factorizations of 6 in A. Of course A is not a PID and the numbers in (3.42) can be expressed as products of non-principal ideals; with an appropriate notion of prime ideal it can be shown that every ring of algebraic integers has unique factorization of ideals (see A.2, Section 9.5).

Exercises 3.6

1. In the ring of Gaussian integers, give a PAQ-reduction for
$$\begin{pmatrix} 1 & 2-\mathrm{i} & 1-\mathrm{i} \\ 1+\mathrm{i} & 6+\mathrm{i} & 2 \end{pmatrix}.$$

2. Determine all units in the ring of Gaussian integers. Do the same for the ring of integers in $\mathbb{Q}(\sqrt{-d})$, where d is any positive square-free integer.

3. Show that the ring of integers generated by the cube roots of 1 is Euclidean.

4. Let R be an integral domain with field of fractions K; an ideal I of R is called **invertible** if there exists an R-submodule J of K such that $IJ = R$. Show that any non-zero principal ideal is invertible; more generally, show that I is invertible if there exists an ideal J such that $IJ = (c)$, where $c \neq 0$.

5. Show that $(2, 1+\sqrt{-5})$ is invertible in the ring of integers of $\mathbb{Q}(\sqrt{-5})$.

6. Give an example of a non-zero ideal in $k[x, y]$ that is not invertible.

4
Ring Constructions

This chapter presents a number of general constructions on rings, such as the direct product and the tensor product (described first for modules) and the basic notions of projective and injective module. Some of the constructions involve the axiom of choice, which is briefly explained, together with the form it usually takes in algebra, viz. Zorn's lemma.

4.1 The Direct Product of Rings

Let R be any ring and suppose that R can be expressed as a direct sum of two right ideals: $R = I \oplus J$. Applying this decomposition to 1, we obtain $1 = e + e'$; for any $x \in I$ we have $x = 1.x = ex + e'x$. It follows that $x - ex = e'x \in I \cap J = 0$, so $e'x = 0$; in particular, $e'e = 0$, or bearing in mind that $e' = 1 - e$, we find that

$$e^2 = e. \tag{4.1}$$

Thus e and e' are idempotent; moreover, $e'e = 0$ and similarly $ee' = 0$. These equations are expressed by saying that e and e' are **orthogonal** idempotents. Conversely, any given idempotent e leads to a decomposition $R = eR + (1 - e)R$, and this sum is direct, for if $x = ey = (1 - e)z$, then $x = 1.x = ex + (1 - e)x = e(1 - e)z + (1 - e)ey = 0$. Thus any representation of R as a direct sum of two right ideals corresponds to an expression of 1 as a sum of two orthogonal idempotents. This result is easily seen to hold for direct sums of any finite number of right ideals. Thus we obtain

Theorem 4.1

Let R be any ring. Then any decomposition of R as a direct sum of right ideals:

$$R = A_1 \oplus A_2 \oplus \ldots \oplus A_r, \tag{4.2}$$

corresponds to a decomposition of 1 as a sum of r pairwise orthogonal idempotents. ■

If the ideals on the right of (4.2) are two-sided, then since each A_i is an ideal, we have $A_iA_j \subseteq A_i \cap A_j = 0$ for $i \neq j$, and it follows that the multiplication in R is determined entirely by the multiplication in each of the A_i. Decomposing the unit element of R as in (4.2), we obtain

$$1 = e_1 + e_2 + \ldots + e_r, \quad e_i \in A_i,$$

where the e_i are pairwise orthogonal idempotents. Given any element $x \in R$, let us decompose x as in (4.2): $x = x_1 + x_2 + \ldots + x_r$; since $1.x = x.1 = x$, we find that $x_i = x_ie_i = e_ix_i$, thus each e_i is central and acts as unit element on A_i. We obtain the following result.

Theorem 4.2

Let R be any ring. Then any decomposition of R as a direct sum of r two-sided ideals corresponds to a decomposition of 1 as a sum of r pairwise orthogonal central idempotents.

It only remains to show that a decomposition of 1 as a sum of pairwise orthogonal central idempotents corresponds to a decomposition of R as a direct sum of two-sided ideals, but this is easily verified, since for any central idempotent e, the set eR is clearly a two-sided ideal. ■

This theorem tells us how to form a ring by taking the direct sum of a finite number of rings. The process will appear a little more natural if we take, not the direct sum, but the direct product.

Given rings $R_1, \ldots, R_n$, the ring R whose additive group is the direct sum of the R_i with componentwise multiplication is called the **direct product** of the R_i and is written

$$R = R_1 \times \ldots \times R_n, \qquad \text{or also } R = \prod_1^n R_i. \tag{4.3}$$

Taking the case of two factors, for simplicity, if $R = R_1 \times R_2$, and the unit element of R_i is $e_i (i = 1, 2)$, then in R, e_1, e_2 are two central orthogonal idempotents whose sum is 1. More generally, in the case of n factors as in (4.3) the unit elements e_i of the R_i are pairwise orthogonal central idempotents whose sum is 1: we have $1 = e_1 + \ldots + e_n$, where 1 is the unit element of R. As in the case of two factors, the R_i are not subrings of R, but they are homomorphic images. We denote the natural projection $R \to R_i$ by φ_i and note the following universal property of the direct product.

Proposition 4.3

Let $R_1, \ldots, R_n$ be any family of rings and denote their direct product by R, with natural projections $\varphi_i : R \to R_i$. Given any ring S with a family of homomorphisms $f_i : S \to R_i$, there exists a unique homomorphism $f : S \to R$ such that $f_i = f\varphi_i$.

Proof

Since the elements of R are n-tuples with the i-th component in R_i, we can define xf for $x \in S$ by describing its components, but if $f\varphi_i$ is to be f_i then the i-th component of xf must be xf_i. This shows that there is just one mapping f such that $f_i = f\varphi_i$, and we see that it is a homomorphism by testing for each component. ■

As an example consider the ring $R = \mathbb{Z}/(6)$. Let us denote the residue class of m by $\underline{m}$; we see that R has two idempotents apart from 0 and 1, namely $\underline{3}$ and $\underline{4}$. Their sum is $\underline{1}$ and they are orthogonal, so they lead to a direct decomposition of R. In fact, we have $\mathbb{Z}/(6) \cong \mathbb{Z}/(2) \times \mathbb{Z}/(3)$. More generally, given any positive integers $m_1, \ldots, m_r$ such that no two have a common factor, briefly if they are **pairwise coprime**, and $m = m_1 \ldots m_r$, then it is the case that

$$\mathbb{Z}/(m) \cong \mathbb{Z}/(m_1) \times \ldots \times \mathbb{Z}/(m_r). \tag{4.4}$$

This result, known as the Chinese remainder theorem (first used by the Chinese calendar makers of the first century AD), is usually stated as

Theorem 4.4

Let $m_1, \ldots, m_r$ be any pairwise coprime integers. Then the congruences

$$x \equiv a_i (\text{mod } m_i) \quad (i = 1, \ldots, r) \tag{4.5}$$

have a common solution, which is unique mod m. The solution is of the form $x = \sum n_i x_i$, where $n_i = m/m_i$ and x_i is a solution of $n_i x_i \equiv a_i \pmod{m_i}$.

Proof

There cannot be more than one solution (mod m), for if both x' and x'' satisfy (4.5), then $x' - x''$ is divisible by each m_i, and since the m_i are pairwise coprime, it is divisible by their product, as asserted. But if x' is one solution, then so is $x' + km$, for all $k \in \mathbb{Z}$, because m is divisible by each m_i. So the solution is at most one congruence class mod m. To find it, let us put $n_i = m/m_i$; then n_i is prime to m_i, therefore we can solve $n_i x_i \equiv a_i \pmod{m_i}$. We claim that $x = \sum n_i x_i$ is a solution of (4.5), for $n_1 x_1$ is a solution for $i = 1$, and adding $n_i x_i$ for $i > 1$ has no effect, because $n_i x_i \equiv 0 \pmod{m_1}$. Thus x satisfies s (4.5) for $i = 1$, and the same argument applies for $i = 2, \ldots, r$. ■

For example, to solve the simultaneous congruences $x \equiv 1 \pmod 3$, $x \equiv 2 \pmod 4$, $x \equiv 3 \pmod 5$, we first solve $20x_1 \equiv 1 \pmod 3$, giving $x_1 \equiv -1$, then $15x_2 \equiv 2 \pmod 4$, giving $x_2 \equiv 2$, and finally $12x_3 \equiv 3 \pmod 5$, giving $x_3 \equiv 4$; hence the solution is $-20 + 30 + 48 \equiv 58$. Alternatively we could replace each of the given congruences by $x \equiv -2$, and read off the solution $x \equiv -2$. Of course such lucky accidents do not normally happen, but we should be on the lookout, to take advantage of them when they do occur.

We remark that in some books the direct product of rings is called the "direct sum"; this term is misleading because, as we saw, the factors are not subrings, even though they have a natural embedding in the product. All that can be said is that the direct product can be written as a direct sum of ideals; this is essentially Theorem 4.2 restated:

Theorem 4.5

Let $R_1, \ldots, R_n$ be any rings and let R be their direct product, with natural projections $\varphi_i : R \to R_i$ and natural embeddings $\psi_i : R_i \to R$. Then $A_i = \text{im}\, \psi_i$ is an ideal in R and we have the direct sum

$$R = A_1 \oplus \ldots \oplus A_n; \tag{4.6}$$

$$\text{where } A_i A_j = 0 \text{ if } i \neq j, \quad A_i A_i \subseteq A_i. \tag{4.7}$$

Conversely, if R can be written as a direct sum of ideals as in (4.6), then the ideals satisfy (4.6) and R is a direct product of rings isomorphic to the A's, regarded as rings.

Proof

The embedding ψ_i is defined by mapping the elements of R_i to the i-th component of the product; then it is clear that the image is an ideal in R and we obtain the direct sum (4.6). Further, we have $A_iA_j \subseteq A_i \cap A_j$, hence (4.7) follows because the sum (4.6) is direct. If e_i is the image of the unit element in A_i, then e_i is a central idempotent, different idempotents are orthogonal and their sum is 1. This argument also shows that if we are given the direct sum (4.6), where the A_i are ideals, then (4.7) holds, and the idempotents can be found as before, and this shows R to be the direct product of the A_i, regarded as rings. ■

We remark that in Proposition 4.3 we need not restrict the number of terms to be finite. Thus let $R_i (i \in I)$ be any family of rings; to avoid ambiguity we shall denote the unit element of R_i by e_i. Then the **direct product** of the R_i is the set-theoretic direct product $\prod R_i$, defined as a ring with the operations performed componentwise: $(x_i) + (y_i) = (s_i)$, $(x_i)(y_i) = (p_i)$, where $s_i = x_i + y_i$, $p_i = x_i.y_i$, with unit element (e_i).

Exercises 4.1

1. Find (i) the least positive integer and (ii) the least integer in absolute value, leaving remainders 2, 3, 2 when divided by 3, 4, 5 respectively.

2. Show that the property of Proposition 4.3 determines the direct product up to isomorphism.

3. Define the direct product R of an arbitrary family of rings (R_i) along the lines of Proposition 4.3 and show that it satisfies the following conditions: if A_i is the kernel of the surjective homomorphism $\varphi : R \rightarrow R_i$ defining R, then (i) $\bigcap A_i = 0$ and (ii) $A_i + A_j = R$ for $i \neq j$ (two ideals, or more generally additive subgroups of R, satisfying (ii) are said to be **comaximal** in R).

4. For any positive integer n write f_n for the natural homomorphism $\mathbb{Z} \rightarrow \mathbb{Z}/(n)$, and write $P = \prod \mathbb{Z}/(n)$ for the direct product of all the $\mathbb{Z}/(n)$, regarded as a ring with the operations performed componentwise, and with projection mappings $\pi_n : P \rightarrow \mathbb{Z}/(n)$. Show that there is an embedding $\varphi : \mathbb{Z} \rightarrow P$ such that $\varphi\pi_n = f_n$.

5. Carry out the same construction as in Exercise 4, (i) with n running over all primes, and (ii) with n running over all powers of 2, and show that both mappings are injective.

6. Show that a 2×2 matrix C over a field is idempotent if and only if $C = I$ or $C = 0$ or det $C = 0$ and tr $C = 1$.

7. Let R be a ring with a finite family of ideals (A_i) and write $R_i = R/A_i$. Verify that the projections $f_i : R \to R_i$ correspond to a homomorphism $f : R \to \prod R_i$. Show that (i) f followed by the projection on R_i is surjective. Further, prove that (ii) f is injective iff $\bigcap A_i = 0$, (iii) f is surjective iff the A_i are pairwise comaximal. If only (i) and (ii) hold, R is said to be a **subdirect product** of the R_i; thus in the above construction, R is a subdirect product of the R_i.

8. Show that an idempotent in a ring R is central iff it commutes with every idempotent of R. (Hint: for any idempotent e and any $x \in R$, consider $e + ex(1 - e)$.)

9. A ring is said to be **reduced**, if $x^2 = 0 \Rightarrow x = 0$. Show that in a reduced ring all idempotents are central.

4.2 The Axiom of Choice and Zorn's Lemma

So far we have mainly had to deal with finite collections of sets and algebras, but since the sets themselves are often infinite, we shall have to handle infinite sets. This raises cardinality questions; we shall keep such questions to a minimum, confining ourselves mainly to **countable** sets, i.e. sets that can be enumerated by means of the natural numbers, and just remark that an example of an uncountable set is given by $\mathbb{R}$, the real numbers (see Exercise 4.2.4). We shall also pass over the question of providing an axiom system for sets, with one exception, the axiom of choice; this states that given any family $\mathscr{F}$ of non-empty sets, we can form a set containing just one member from each set of $\mathscr{F}$. At first sight this sounds self-evident, but the need for such an axiom can be justified intuitively as follows: if we have a construction involving no choices, we can usually describe the process by a formula, and this is valid even if the construction is repeated an infinite number of times. However, if our construction involves a choice, then there will be no formula for carrying out an infinite number of choices, and the axiom is needed. The point was well illustrated by Bertrand Russell (1872–1970), who tells the story of a millionaire who has infinitely many pairs of shoes and infinitely many pairs of socks. To pick one shoe from each pair is easy: he can just pick the left shoe each time, but to pick one sock from each pair he needs the axiom of choice.

The axiom of choice was first used by Ernst Zermelo (1871–1953) in his proof (in 1904) that every set can be well-ordered, where a set is well-ordered if it has an ordering such that every non-empty subset has a least element. We shall not give a proof, but merely note the converse result: if every set can be well-ordered, and we are given any family of non-empty sets, we can choose one member from each set by well-ordering the union of all the sets and from each set take the least element in the well-ordering.

There is another form of the axiom of choice which is particularly useful in algebra, and this is the form we shall use. It is a maximal principle known as **Zorn's lemma** (after Max Zorn, 1906–1993). To state it we shall say that a partially ordered set S is **inductive** if every chain in S has an upper bound in S. We observe that such a set cannot be empty, since the empty chain must have an upper bound.

Zorn's Lemma

Every partially ordered set which is inductive, has an upper bound.

The proof requires the axiom of choice or an equivalent condition; we shall omit it and assume it as an axiom (for a proof, see e.g. Kaplansky, 1972). In applying Zorn's lemma it is important to check that every chain is inductive, and not merely every ascending sequence. To give an example, consider the set $\mathscr{C}$ of all countable sets of real numbers under ordering defined by set inclusion. Every countable sequence of members of $\mathscr{C}$ has an upper bound, because a countable sequence of countable sets is itself countable, but $\mathscr{C}$ has no maximal element, for since $\mathbb{R}$ is uncountable, no countable subset can be maximal.

Zorn's lemma has many applications in ring theory and in algebra generally; as an illustration we shall prove a result on modules that is frequently used. We recall that a **maximal proper** submodule N of a module M is a submodule N that is maximal among the proper submodules, i.e. $N \subset M$ and $N \subseteq M' \subset M$ implies $M' = N$.

Theorem 4.6

Let R be any ring and M a finitely generated R-module. Then every proper submodule of M is contained in a maximal proper submodule.

Proof

Let N be a proper submodule of M. We take the family of all proper submodules of M containing N and show that it is inductive. Thus, given a chain $\{P_\lambda\}$ of proper submodules, where $P_\lambda \supseteq N$, we have to find an upper bound, i.e. a proper submodule containing all the P_λ. For convenience we order the suffixes by writing $\lambda \leq \mu$ if $P_\lambda \subseteq P_\mu$. Let $P = \bigcup P_\lambda$; we claim that P is proper. For if not, and if $\{u_1, \ldots, u_n\}$ is a finite generating set of M, suppose that $u_i \in P_{\lambda_i}$ for $i = 1, \ldots, n$. Among the suffixes $\lambda_1, \ldots, \lambda_n$ there is a largest, say $\mu = \lambda_i$; then $P_\mu \ni u_1, \ldots, u_n$ and so $P_\mu = M$, which is a contradiction. It follows that P is proper; thus our chain has an upper bound and so the family of proper submodules containing N is inductive, and by Zorn's lemma, there is a maximal proper submodule containing N. ■

Here the hypothesis of finite generation was essential; e.g. the additive group of rational numbers, $\mathbb{Q}$, is not finitely generated and it has no maximal proper subgroups (see Exercise 4.2.10). As an important consequence we have **Krull's theorem**:

Corollary 4.7

Let R be a ring. Then every proper right ideal is contained in a maximal proper right ideal. The same is true for left or two-sided ideals.

This follows from Theorem 4.6, because R, as right R-module, is generated by a single element, namely 1; likewise for R as left R-module or as R-bimodule. ■

Exercises 4.2

1. Show that Zorn's lemma is equivalent to Hausdorff's maximal principle (Felix Hausdorff, 1868–1942): in any partially ordered set every chain is contained in a maximal chain. (Hint: apply Zorn's lemma to the set of chains containing a given one.)

2. Show that the union of countably many countable sets is itself countable. (Hint: enumerate the sets as rows and enumerate the union by going along diagonals.)

3. Show that any subset of a countable set is again countable. Use this fact and the result of Exercise 2 to show that the set of all rational numbers is countable.

4. Show that the set of all real numbers between 0 and 1 is uncountable. (Hint: if an enumeration exists, write down decimals for each number and try to construct a decimal different from all these. This is known as **Cantor's diagonal method** (Georg Cantor, 1845–1918); care must be taken because some numbers have more than one decimal expansion, e.g. $0.5 = 0.4999\dots$.)

5. The axiom of choice is sometimes stated as the **multiplicative axiom**: the product of any family of non-empty sets is non-empty. Prove their equivalence.

6. Show without using the axiom of choice or the multiplicative axiom that the product of any family of finite non-empty sets of real numbers is non-empty. Do the same for the product of any family of non-empty subsets of $\mathbb{Z}$.

7. Show (using Zorn's lemma) that every vector space over a field (even skew) has a basis.

8. Which two of the four implications of Theorem 2.2 use the axiom of choice?

9. Let R be a ring and A a commutative subring. Show that there is a maximal commutative subring of R containing A.

10. Show that $\mathbb{Q}$, as additive group, has no maximal proper subgroups.

4.3 Tensor Products of Modules and Algebras

In the study of vector spaces over a field k the **bilinear** mappings play an important role; thus we are given three vector spaces U, V and W over k and a mapping $f(x, y)$ from $U \times V$ to W such that $f(u, y)$ for each fixed $u \in U$ is linear in y and $f(x, v)$ for each fixed $v \in V$ is linear in x. If U and V have bases $u_1, \dots, u_m$ and $v_1, \dots, v_n$ respectively, then the value of $f(x, y)$ can be expressed as a k-linear combination of $f(u_i, v_j)$, for $i = 1, \dots, m$, $j = 1, \dots, n$. Formally, f may be described as a linear function on a "product space" $U \otimes V$ with basis $u_i \otimes v_j$. This space $U \otimes V$ is called the **tensor product** of U and V; let us see how it can be defined for modules over general rings. Since these modules in general no longer have bases, we have to proceed differently; to simplify matters we shall for the present assume our ring to be commutative and leave the general case to be dealt with in Section 4.4.

Thus let K be a commutative ring and let U, V, W be any K-modules; we shall take them to be left K-modules, although as we have seen, this is only a matter of notation. We want to consider bilinear mappings from U, V to W, i.e. mappings

$$f : U \times V \to W, \tag{4.8}$$

such that f is K-linear in each argument. We claim that there is a K-module T with a bilinear mapping $\lambda : U \times V \to T$ which is universal for bilinear mappings (4.8); in detail this means that to any bilinear mapping f as in (4.8) there corresponds a unique homomorphism $f' : T \to W$ such that the triangle below is commutative:

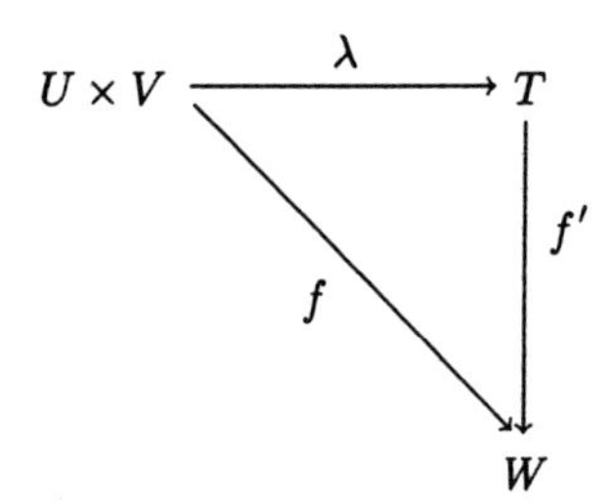

Like every universal object, T is unique up to isomorphism, provided that it exists. To prove its existence we take the set X of all pairs (u, v), where $u \in U, v \in V$, and form the free K-module A on this set X. Let B be the submodule of A generated by all the elements

$$\left.\begin{array}{rl} (u+u', v) - (u, v) - (u', v) & u, u' \in U \\ (u, v+v') - (u, v) - (u, v') & v, v' \in V, \\ (\alpha u, v) - \alpha(u, v), (u, \alpha v) - \alpha(u, v) & \alpha \in K. \end{array}\right\} \tag{4.9}$$

Consider the inclusion mapping $\iota : U \times V \to A$ and define $\lambda : U \times V \to A/B$ as the mapping ι followed by the natural homomorphism from A to A/B. This mapping is bilinear, for the elements (4.9) generating B were just chosen to ensure this. We now put $T = A/B$ and claim that T with the mapping λ is the required universal object. Given any bilinear mapping $f : U \times V \to W$, ignoring the bilinearity, i.e. as a mapping of sets, it extends to a unique homomorphism $f_1 : A \to W$, since A is free on the (u, v). We claim that $\ker f_1 \supseteq B$; for we have

$$[(u+u', v) - (u, v) - (u', v)]f_1 = (u+u', v)f - (u, v)f - (u', v)f = 0,$$
$$[(\alpha u, v) - \alpha(u, v)]f_1 = (\alpha u, v)f - \alpha(u, v)f = 0,$$

and similarly for the remaining relation (4.9), by the bilinearity of f. By the factor theorem it follows that f_1 induces a homomorphism $\lambda : A/B = T \to W$, and this has the required properties, by its construction. It is unique, since its values are determined by the images of the (u, v), which form a generating set of T. The module T is called the **tensor product** of U and V over K and is denoted by $U \otimes_K V$, or by $U \otimes V$ when there is no risk of confusion. We state the result as follows.

Theorem 4.8

Let K be a commutative ring. Then for any K-modules U, V there exists a K-module $U \otimes_K V$ together with a bilinear mapping $\lambda : U \times V \to U \otimes_K V$ which is universal for bilinear mappings from $U \times V$ to K-modules. ■

The image of (u, v) in $U \otimes V$ is denoted by $u \otimes v$; thus $U \otimes V$ is a K-module with generating set $\{u \otimes v | u \in U, v \in V\}$ and defining relations

$$(u + u') \otimes v = u \otimes v + u' \otimes v, \quad u, u' \in U,$$
$$u \otimes (v + v') = u \otimes v + u \otimes v', \quad v, v' \in V,$$
$$(\alpha u) \otimes v = u \otimes (\alpha v) = \alpha(u \otimes v), \quad a \in K.$$

For example, if K is a field and U, V are vector spaces with bases $u_1, \ldots, u_m$, $v_1, \ldots, v_n$, respectively, then it is easily verified that $U \otimes V$ has the basis $u_i \otimes v_j \, (i = 1, \ldots, m, j = 1, \ldots, n)$; in particular, we note that in this case we have the following relation between the dimensions of these spaces:

$$\dim(U \otimes V) = (\dim U)(\dim V).$$

For modules over general rings we cannot expect such a relation because modules do not in general have a dimension. Tensor products over general rings behave in ways that are at first unexpected, but they form an important construction with many uses. To give just one example, let us take $\mathbb{Z}$-modules, i.e. abelian groups. We shall put $A = \mathbb{Z}$, $C = \mathbb{Z}/(2)$ and consider the exact sequence

$$0 \to A \xrightarrow{\varphi} A \xrightarrow{\nu} C \to 0.$$

where $\varphi : x \mapsto 2x$ and ν is the natural homomorphism. If we tensor with C and bear in mind that $A \otimes C \cong C$, $C \otimes C \cong C$ (because $C = \{0, 1\}$ and the only non-zero element of the form $a \otimes b, a, b \in C$ is $1 \otimes 1$), we obtain the sequence

$$0 \to C \to C \to C \to 0,$$

which is clearly not exact; in fact it fails to be exact at the first C. This is in contrast to the situation for vector spaces over a field, where tensoring with a vector space preserves exactness.

Even for vector spaces over a field it is not always convenient to choose bases for the two factors. Here it is useful to have the following **independence rule**.

Theorem 4.9

Let k be a field and U, V any vector spaces over k. Given linearly independent vectors $u_1, \ldots, u_r \in U$, if for some $y_1, \ldots, y_r \in V$,

$$\sum u_i \otimes y_i = 0 \quad \text{in } U \otimes V,$$

then $y_1 = \ldots = y_r = 0$.

The proof is almost immediate, by completing $u_1, \ldots, u_r$ to a basis of U. ■

We note that this result does not generalize to modules over more general rings, as the previous example shows. However, it is an easy consequence of linearity that if U is generated by u_i $(i = 1, \ldots, r)$ and V by v_j $(j = 1, \ldots, s)$ (over any K), then $U \otimes V$ is generated by $u_i \otimes v_j$ $(i = 1, \ldots, r, j = 1, \ldots, s)$ (see also Exercises 4.3.9 and 4.3.10).

To give another example, we saw (in Exercise 8 of Section 1.5) that the direct sum of two cyclic groups of coprime orders m, n is cyclic of order mn. By contrast their tensor product is zero; we shall verify this in the case $m = 2$, $n = 3$ and leave the general case as an exercise. Thus let $A = \{0, 1\}$, $B = \{0, 1, 2\}$ be cyclic of orders 2, 3, respectively, and consider $A \otimes B$; it is generated by the element $c = 1 \otimes 1$ and we have $2c = 2 \otimes 1 = 0 \otimes 1 = 0$, $3c = 1 \otimes 3 = 0$, hence $c = 0$. The only other element is $1 \otimes 2$; clearly it equals $2 \otimes 1 = 0$, and this shows that $A \otimes B = 0$.

Let U, V be vector spaces over k, with bases $u_1, \ldots, u_r$ and $v_1, \ldots, v_s$ respectively and suppose we have linear endomorphisms of U and V with matrices A, B referred to these bases. Then the mapping $u_i \otimes v_k \mapsto \sum a_{ij} b_{kl} u_j \otimes v_l$ defines an endomorphism of $U \otimes V$; the corresponding matrix is denoted by $A \otimes B$; it is an $rs \times rs$ matrix, which the reader may wish to write out in simple cases. We shall not do so but merely note the trace formula

$$\operatorname{tr}(A \otimes B) = \sum a_{ii} b_{kk} = \left(\sum a_{ii}\right)\left(\sum b_{kk}\right) = (\operatorname{tr} A)(\operatorname{tr} B). \tag{4.10}$$

We now take a brief look at tensor products of K-algebras, where K is a commutative ring. Any K-algebra A is by definition a K-module with a mapping

$\lambda_0 : A \times A \to A$ which is bilinear. By the definition of the tensor product of modules this is equivalent to prescribing a linear mapping

$$\lambda : A \otimes A \to A; \tag{4.11}$$

this mapping λ is also called the **multiplication** in A. Explicitly we have

$$\Big(\sum x_i \otimes y_i\Big)\lambda = \sum x_i y_i, \quad \text{for } x_i, y_i \in A.$$

In order to introduce the tensor product of algebras we shall need the transposition mapping. Given any two K-modules U, V, we define the **transposition mapping** $\tau : V \otimes U \to U \otimes V$ by the rule

$$\tau : y \otimes x \mapsto x \otimes y. \tag{4.12}$$

Let A and B be any K-algebras, with respective multiplications λ and μ, and let τ be the transposition interchanging factors from B and A as in (4.12). It gives rise to a permutation mapping $\tau_1 = 1 \otimes \tau \otimes 1 : A \otimes B \otimes A \otimes B \to A \otimes A \otimes B \otimes B$, where

$$(a_1 \otimes b_1 \otimes a_2 \otimes b_2)\tau_1 = a_1 \otimes a_2 \otimes b_1 \otimes b_2. \tag{4.13}$$

By combining τ_1 with the multiplications λ of A and μ of B we obtain a linear mapping

$$\nu = \tau_1(\lambda \otimes \mu) : A \otimes B \otimes A \otimes B \to A \otimes B. \tag{4.14}$$

We claim that this multiplication is associative whenever λ and μ are. Writing $C = A \otimes B$, we can express (4.14) as $\nu : C \otimes C \to C$, and now we obtain for any $a_i \in A, b_i \in B$,

$$\begin{aligned}(a_1 \otimes b_1 \otimes a_2 \otimes b_2 \otimes a_3 \otimes b_3)(\nu \otimes 1)\nu &= (a_1 a_2 \otimes b_1 b_2 \otimes a_3 \otimes b_3)\nu \\ &= (a_1 a_2)a_3 \otimes (b_1 b_2)b_3.\end{aligned}$$

Applying $(1 \otimes \nu)\nu$, we obtain $a_1(a_2 a_3) \otimes b_1(b_2 b_3)$, which is the same, by the associativity in A and B. Since the elements on the left span $C \otimes C \otimes C$, the associativity of ν follows. Similarly, if A and B are commutative, it follows that C is commutative. When A has a unit element e, then the mapping

$$b \mapsto e \otimes b \tag{4.15}$$

is a homomorphism from B to $A \otimes B$, and likewise a unit element in B leads to a homomorphism $A \to A \otimes B$. If A, B both have unit elements e, f, respectively, then $e \otimes f$ is a unit element for C, as is easily checked. We sum up the result as

Theorem 4.10

Let K be a commutative ring and A, B any K-algebras. Then the tensor product $A \otimes B$ is again a K-algebra, which is associative, commutative or has a unit element whenever this is so for A and B. ■

We have already seen that if A, B are algebras over a field with respective bases $\{u_i\}$, $\{v_j\}$ then $A \otimes B$ is an algebra with basis $\{u_i \otimes v_j\}$. Here the bases are usually finite; this is certainly the only case we shall be concerned with, but in fact the result holds quite generally, although in the infinite-dimensional case one may want to form a completion. Let us mention some other examples that we shall meet again later.

1. Let K be a field and E an extension field. Given any K-algebra A, with K-basis $u_1, \dots, u_r$, say, we can form the tensor product $A \otimes E$, which is an algebra over E with basis $u_1 \otimes 1, \dots, u_r \otimes 1$. Regarded as E-algebra, $A \otimes E$ is usually denoted by A_E and is called the algebra obtained from A **by extension of the ground field** to E.
2. For any commutative ring K, the full matrix ring over K, $\mathfrak{M}_r(K)$, has as basis over K the matrix units $e_{ij}(i, j = 1, \dots, r)$ with the multiplication rule $e_{ij}e_{kl} = \delta_{jk}e_{il}$. Given any K-algebra A, we can express the full matrix ring over A, $\mathfrak{M}_r(A)$, as a tensor product
$$\mathfrak{M}_r(A) = \mathfrak{M}_r(K) \otimes A. \tag{4.16}$$
In the special case where $A = \mathfrak{M}_s(K)$, the left-hand side of (4.16) is $\mathfrak{M}_r(\mathfrak{M}_s(K)) \cong M_{rs}(K)$, for the elements of $\mathfrak{M}_r(\mathfrak{M}_s(K))$ are $r \times r$ matrices whose entries are $s \times s$ matrices over K. We thus obtain the isomorphism
$$\mathfrak{M}_r(K) \otimes \mathfrak{M}_s(K) \cong \mathfrak{M}_{rs}(K). \tag{4.17}$$

For a given K-algebra we may wish to know whether it can be expressed as a tensor product of certain subalgebras. Here the following criterion is useful. Given an algebra C over a field K and two submodules U, V of C, we shall say that U and V are **linearly disjoint** over K, if for any linearly independent elements $\{u_i\}$ in U and $\{v_j\}$ in V the elements u_iv_j in C are linearly independent over K. This just means that the natural mapping $U \otimes V \to C$ induced by the mapping $(x, y) \mapsto xy$ is injective. We now have the following criterion for decomposability as a tensor product.

Theorem 4.11

Let C be an algebra over a field K. If C has subalgebras A and B such that the mapping

$$(x, y) \mapsto xy \tag{4.18}$$

defines an isomorphism

$$A \otimes B \cong C, \tag{4.19}$$

then (i) A and B commute elementwise, (ii) $AB = C$ and (iii) A and B are linearly disjoint in C. Conversely, when (i)–(iii) hold, then (4.19) holds, with the mapping induced by (4.18).

Proof

The verification of (i)–(iii) when (4.19) holds is straightforward and may be left to the reader. For the converse, the mapping (4.18) from $A \times B$ to C is bilinear and so induces a K-linear mapping $A \otimes B \to C$. This mapping is a homomorphism by (i), surjective by (ii) and injective by (iii); hence it is an isomorphism. ■

Exercises 4.3

1. If a module M is a direct sum: $M = M' \oplus M''$, show that $M \otimes N \cong (M' \otimes N) \oplus (M'' \otimes N)$.
2. Show that if A, B are two cyclic groups of coprime orders, then their tensor product is zero. What is the tensor product $A \otimes A$, or of $(A \oplus B) \otimes (A \oplus B)$?
3. Given three modules U, V, W over a commutative ring K, show that $\mathrm{Hom}(U, \mathrm{Hom}(V, W)) \cong \mathrm{Hom}(U \otimes V, W)$.
4. Let R be a commutative integral domain with field of fractions K. For any R-module M define M_K as $M \otimes_R K$ and show that the mapping $x \mapsto x \otimes 1$ is a homomorphism from M to M_K which is injective iff M is torsion-free.
5. With R, K, M, M_K as in Exercise 4 define the **rank** of M, $\mathrm{rk}(M)$ as the dimension of M_K as K-space. Show that the functor associating M_K with M is exact (as defined in Section 4.4 below); deduce that for any short exact sequence $0 \to M' \to M \to M'' \to 0$, $\mathrm{rk}(M) = \mathrm{rk}(M') + \mathrm{rk}(M'')$.

6. Show that for a tensor product of square matrices A, B over a field, the determinant satisfies the formula $\det(A \otimes B) = (\det A)(\det B)$.

7. Let A, B be any (unital) k-algebras, not assumed associative. Show that the mapping from A to $A \otimes B$ defined by $x \mapsto x \otimes 1$ is a homomorphism. If the tensor product $A \otimes B$ is associative, show that A and B are also associative.

8. Let A, B be k-algebras, with subspaces X, Y respectively, and denote their centralizers by X', Y'. Show that the centralizer of $X \otimes Y$ in $A \otimes B$ is $X' \otimes Y'$.

9. For any ring R (not necessarily commutative) let U be a right R-module and V a left R-module. Given a relation

$$\sum_{i=1}^{n} x_i \otimes y_i = 0 \tag{4.20}$$

in $U \otimes V$, let V' be the submodule generated by $y_1, \ldots, y_n$. Show that (4.20) need not hold in $U \otimes V'$.

10. Let R, U, V be as in Exercise 9 and assume further that every finitely generated submodule of V is contained in a free submodule. If a relation (4.20) holds in $U \otimes V$, where $y_1, \ldots, y_n$ are left linearly independent, show that $x_1 = \ldots = x_n = 0$.

4.4 Modules over General Rings

In the definition of the tensor product of modules in Section 4.3 we assumed that the ground ring is commutative, and at first sight it seems difficult to define the tensor product of two left R-modules when R is non-commutative. We meet a similar problem when we try to define an R-module structure on $\mathrm{Hom}_R(M, N)$. These problems are easily resolved by regarding all our modules as bimodules; we recall from Section 1.4 that for any rings R, S, an R-S-bimodule is a defined as a left R-module M which is also a right S-module such that $(rm)s = r(ms)$ for all $m \in M, r \in R, s \in S$. When $S = R$, we speak of an R-bimodule, while a left R-module may always be considered as an R-$\mathbb{Z}$-bimodule.

Let us now consider $H = \mathrm{Hom}_R(M, N)$, where M is an R-S-bimodule and N is an R-T-bimodule. More briefly, we shall say that we have the situation $({}_RM_S, {}_RN_T)$. We know that H is an abelian group and we ask whether it also has a module structure. If we write the effect of $f \in H$ on $x \in M$ as (x, f), the R-linearity of f is expressed by the equation

$$(rx, f) = r(x, f), \quad \text{where } x \in M, r \in R, f \in H. \tag{4.21}$$

We can define H as a right T-module by the equation

$$(x, ft) = (x, f)t, \quad \text{where } x \in M, f \in H, t \in T. \tag{4.22}$$

To check the module condition, we have, for $t, t' \in T$, $(x, f(tt')) = (x, f)(tt') = ((x, f)t)t' = (x, ft)t' = (x, (ft)t')$, hence $f(tt') = (ft)t'$. A natural question is whether we can also define H as a right S-module; this leads to difficulties (unless S is commutative), but it can be defined as a **left** S-module, by the rule

$$(x, sf) = (xs, f), \quad \text{where } x \in M, f \in H, s \in S. \tag{4.23}$$

Again the module condition is easily verified; details are left to the reader. Moreover, H in this way becomes an S-T-bimodule, for we have $(x, s(ft)) = (xs, ft) = (xs, f)t = (x, sf)t = (x, (sf)t)$, and so $s(ft) = (sf)t$. The result may be summed up as

Theorem 4.12

Given any rings R, S, T, an R-S-bimodule M and an R-T-bimodule N, the set $\mathrm{Hom}_R(M, N)$ has an S-T-bimodule structure defined by (4.22), (4.23). ∎

This reversal of the action of S may be explained from the point of view of category theory. The functor Hom(M, -) is covariant, therefore it preserves the T-action, whereas the functor Hom(-, N) is contravariant (cf. Section 1.5) and so it reverses the S-action: to a right S-module M corresponds a left S-module $\mathrm{Hom}_R(M, N)$.

As an illustration of Theorem 4.12 let us take $M = R$. As left R-module, R is generated by 1, and so any $f \in \mathrm{Hom}(R, N)$ is completely determined by its value on 1. If $1f = x \in N$, then $af = (a.1)f = a(1.f) = ax$; it follows that $\mathrm{Hom}(R, N) \cong N$. Clearly the result holds for right as well as for left modules, so we obtain

Corollary 4.13

For any ring R and any R-module N, $\mathrm{Hom}_R(R, N) \cong N$. ∎

Another important example illustrating Theorem 4.12 is obtained by taking $N = M$ (and $T = S$). The set $\mathrm{Hom}_R(M, M)$ consists of all endomorphisms of M and is usually written $\mathrm{End}_R(M)$. It admits a multiplication, which is associative and distributive over addition and so defines a ring structure on $\mathrm{End}_R(M)$, which thus becomes the **endomorphism ring** of M. From Theorem 4.12 we see that when M is an R-S-bimodule, then $\mathrm{End}_R(M)$ is an S-bimodule.

In Section 2.1 we defined a k-algebra; the definition may be summed up by saying that a k-algebra is a ring R with a homomorphism of k into the centre of R. In this form the definition makes sense for any ring k, not necessarily a field, or even a commutative ring, although the restriction for the image to lie in the centre of R seems now a little artificial. To overcome this limitation, we define, for any ring K, a K-**ring** as a ring R which is a K-bimodule such that

$$\alpha(rs) = (\alpha r)s, \quad (r\alpha)s = r(\alpha s), \quad (rs)\alpha = r(s\alpha), \quad \text{for } r, s \in R, \alpha \in K. \tag{4.24}$$

Clearly the definition can again be expressed by saying that a K-ring is just a ring R with a homomorphism of K into R, but the more explicit form of the definition can still be used when R does not have a unit element. As an example of a K-ring, given an R-K-bimodule M, we may regard $\mathrm{End}_R(M)$ as a K-ring.

For any ring K and any group G we can form the free K-module on G as basis, with multiplication as in G. The resulting ring KG is called the **group ring**; clearly it is a K -ring which generalizes the notion of group algebra.

We now consider the tensor product. We recall from Section 4.3 that it is defined as a universal object for bilinear mappings, but the problem is how to define bilinear mappings from a pair of left R-modules to a left R-module. The answer is again to take bimodules, but this time we shall take an S-R-bimodule U, an R-T-bimodule V and an S-T-bimodule W; thus we have the situation $({}_SU_R, {}_RV_T, {}_SW_T)$. Then we can speak of S-T-bilinear mappings, defined as mappings $\varphi : U \times V \to W$ which are **biadditive**, i.e. additive in each argument and satisfy

$$(su, v)\varphi = s(u, v)\varphi, \quad (u, vt)\varphi = (u, v)\varphi t \quad \text{for all } u \in U, v \in V, s \in S, t \in T. \tag{4.25}$$

We shall further require φ to be **balanced**, i.e. such that

$$(ur, v)\varphi = (u, rv)\varphi \quad \text{for all } u \in U, v \in V, r \in R. \tag{4.26}$$

Of course it is necessary to check that these definitions make sense, e.g. for (4.26) this follows because for any $r, r' \in R$, $(u(rr'), v)\varphi = ((ur)r', v)\varphi = (ur, r'v)\varphi = (u, r(r'v))\varphi = (u, (rr')v)\varphi$. It is instructive to work out why a similar verification for bilinearity fails when U, V are both left R-modules or both right R-modules, and the reader is urged to do so. We can now as before define an abelian group which is a universal object for balanced bilinear mappings from $U \times V$ to W; it is denoted by $U \otimes_R V$ or simply $U \otimes V$, when there is no risk of confusion, and it can be defined as a left S-module and right T-module by putting

$$s(u \otimes v) = su \otimes v, \quad (u \otimes v)t = u \otimes vt, \quad \text{for } u \in U, v \in V, s \in S, t \in T.$$

Further, with these definitions $U \otimes V$ becomes an S-T-bimodule, as is easily checked.

Of course the balanced biadditive mappings from $R \times V$ to W are just the T-linear mappings from V to W, because $(r, v)\varphi = (1, rv)\varphi$. It follows that $R \otimes V \cong V$, and similarly $U \otimes R \cong U$.

Now let Q, R, S, T be any rings and consider the situation $({}_QU_R, {}_RV_S, {}_TW_S)$. We can form $\mathrm{Hom}_S(V, W)$, the set of S-linear mappings from V to W; as we have seen, this is a T-R-bimodule. So we can form $\mathrm{Hom}_R(U, \mathrm{Hom}_S(V, W))$, which is a T-Q-bimodule; it consists of all mappings from $U \times V$ to W which are biadditive and preserve the action by S. Moreover, these mappings are R-balanced, as is easily checked. But these are just the S-linear mappings from $U \otimes V$ to W, so we have an isomorphism

$$\mathrm{Hom}_R(U, \mathrm{Hom}_S(V, W)) \cong \mathrm{Hom}_S(U \otimes_R V, W). \tag{4.27}$$

Moreover, this isomorphism is natural, in that it does not depend on a choice of bases in U, V and W. It is known as **adjoint associativity**. More generally, suppose we are given two functors $S : \mathscr{A} \to \mathscr{B}$, $T : \mathscr{B} \to \mathscr{A}$, where $\mathscr{A}$, $\mathscr{B}$ are categories whose hom-sets are abelian groups (e.g. module categories); S is called a **left adjoint** and T a **right adjoint** and S, T are an **adjoint pair** if for any $\mathscr{A}$-object X and $\mathscr{B}$-object Y,

$$\mathscr{B}(X^S, Y) \cong \mathscr{A}(X, Y^T), \tag{4.28}$$

where $\cong$ indicates a natural isomorphism of abelian groups. Thus (4.27) expresses the fact that the tensor product is the left adjoint of the covariant hom functor. To give another example, let F_X be the free group on a set X and for any group G denote by G^U the underlying set (the forgetful functor from groups to sets). These functors form an adjoint pair, for on writing Gp for the category of groups and homomorphisms, we have

$$\mathrm{Gp}(F_X, G) \cong \mathrm{Ens}(X, G^U).$$

Generally most universal constructions are left adjoints of forgetful functors. Anticipating a definition from Section 4.6, we observe that every left adjoint functor is right exact and every right adjoint functor is left exact (for a proof, see e.g. A.3, Chapter 3); in particular this shows the covariant hom functor $\mathrm{Hom}(U, \text{-})$ to be left exact and the tensor product functor $U \otimes$ - to be right exact, by (4.27).

We conclude this section by describing the radical of a general ring. We recall from Theorem 2.24 that the radical of an Artinian ring R may be defined as the intersection of all maximal right (or equivalently, left) ideals of R. We first note some general properties.

Theorem 4.14

Given any ring R and any element $a \in R$, the following properties are equivalent:

(a) a belongs to each maximal left ideal of R;

(b) $1 - xa$ has a left inverse for each $x \in R$;

(c) $1 - xay$ has an inverse for all $x, y \in R$;

(d) a annihilates every simple left R-module;

(a°)–(d°) the right-hand analogues of (a)–(d).

Proof

(a) $\Rightarrow$ (b): we shall prove that not(b) implies not(a). Assume that $1 - xa$ has no left inverse. Then $R(1 - xa)$ does not contain 1 and so is a proper left ideal; by Krull's theorem (Corollary 4.7) it is contained in a maximal left ideal P. If P were to contain a, it would also contain $xa + (1 - xa) = 1$, a contradiction, hence $a \notin R(1 - xa)$.

We next prove (b) $\Rightarrow$ (d). Let M be a simple left R-module; we have to show that $au = 0$ for all $u \in M$, so assume that $au \neq 0$. Then $Rau = M$ by simplicity, hence $u = xau$ for some $x \in R$; hence $(1 - xa)u = 0$, but $1 - xa$ has a left inverse z, say, so $u = z(1 - xa)u = 0$, which contradicts the fact that $au \neq 0$. Hence $aM = 0$.

Clearly (d) $\Rightarrow$ (a), for if (d) holds and P is a maximal left ideal, then R/P is a simple left R-module, hence $a(R/P) = 0$, i.e. $aR \subseteq P$, and so $a \in P$.

Thus (a), (b), (d) are equivalent; further (c) evidently implies (b); we complete the proof by showing that (b) and (d) together imply (c). By (d), $aM = 0$ for every simple left R-module M, hence $ayM \subseteq aM = 0$ for any $y \in R$; hence ay satisfies (d) and hence (b). Therefore $1 - xay$ has a left inverse, which we shall write as $1 - b$:

$$(1 - b)(1 - xay) = 1. \tag{4.29}$$

By writing this as $b = (b - 1)xay$, we see that $bM = 0$, hence $1 - b$ has a left inverse $1 - c$:

$$(1 - c)(1 - b) = 1. \tag{4.30}$$

Hence $1 - xay = (1-c)(1-b)(1-xay) = 1-c$, and this shows $1-b$ to be a two-sided inverse of $1 - xay$.

This shows (a)–(d) to be equivalent, and the evident symmetry of (c) shows their equivalence to (a°)–(d°). ■

We remark that $1-a$ and $1-a'$ are inverse iff the equations

$$a + a' = aa' = a'a \tag{4.31}$$

hold; like the inverse, if a' exists, it is uniquely determined by a. It is called the **quasi-inverse** of a. Now the set $\mathscr{J}(R)$ of all $a \in R$ satisfying the conditions of Theorem 4.14 is called the **Jacobson radical** of R (after Nathan Jacobson, 1910–1999). By Condition (a) of Theorem 4.14 and Theorem 2.24 $\mathscr{J}(R)$ reduces to the usual radical when R is Artinian. Of course for general rings this radical need not be nilpotent, but the form of Nakayama's lemma proved in Section 2.4 still holds, with the same proof.

Many of the other properties of the radical of an Artinian ring are not shared by the Jacobson radical of a general ring, thus $\mathscr{J}(R)$ is in general not a nilideal, nor is the quotient semisimple (for precise conditions see H. Bass, 1960), but the ring $S = R/\mathscr{J}(R)$ has the property $\mathscr{J}(S) = 0$; such a ring is said to be **semiprimitive**. Of course for Artinian rings this notion reduces to semisimplicity; in general $\mathscr{J}(R)$ may be described as the maximal ideal of R with semiprimitive quotient, but a detailed account of these matters would go beyond the framework of this book (cf. A.3, Chapter 10).

Exercises 4.4

1. Let N be an R-T-bimodule. Verify that $\mathrm{Hom}_R(\text{-}, N)$, operating on R-S-bimodules, is an S-T-bimodule. Show also why it fails to be an R-module for a non-commutative ring R.

2. Prove the following form of Nakayama's lemma for a general ring R: If M is a finitely generated left R-module such that $\mathscr{J}(R)M = M$, then $M = 0$.

3. Deduce Nakayama's lemma in the form proved in Section 2.4 from Exercise 2.

4. Find $\mathscr{J}(R)$, where $R = k[[x]]$ is a formal power series ring over a field k and show that it is not nilpotent.

5. Show that for any ring R and any positive integer n, $\mathscr{J}(R_n) = \mathscr{J}(R)_n$.

6. Let R be any ring and A an ideal in R. Show that if $\mathscr{J}(R/A) = 0$ holds, then $\mathscr{J}(R) \subseteq A$.

7 Let R be any ring and A an ideal in R such that $A \subseteq \mathscr{J}(R)$. Show that $\mathscr{J}(R/A) \cong \mathscr{J}(R)/A$.

8. If R is a ring with no non-zero nilideals, show that the polynomial ring $R[t]$ has zero radical. (Hint: for any $f \in \mathscr{J}(R[t])$, show that the existence of a quasi-inverse for ft leads to a contradiction.)

4.5 Projective Modules

Free modules play an important role in the study of rings; they are easily defined, but the definition makes reference to a basis of the module, even though this basis is far from unique. We shall now meet a somewhat larger class, the projective modules; they have the advantage of being defined in a more invariant fashion than the free modules, which in many situations makes them even more useful than free modules, although they are not as easily described.

We begin by noting a property of free modules. Given a short exact sequence

$$0 \to B \xrightarrow{f} A \xrightarrow{g} C \to 0, \tag{4.32}$$

if C is free, then this sequence is split exact. To see this we choose a basis $\{u_i\}$ of C and for each u_i pick $x_i \in A$ such that g maps x_i to u_i. Since C is free on the u_i, the mapping $u_i \mapsto x_i$ can be extended to a homomorphism $\gamma : C \to A$, and since $\gamma g = 1$, the sequence (4.32) is split in this way.

We now define a module P to be **projective** if any short exact sequence with P as third term is split exact. The previous remark shows that every free module is projective; in general there will be other projective modules. To investigate this question more closely, we shall need to look at functors; we recall from Section 1.4 that they are the mappings between categories that preserve composition. At first we shall mainly be concerned with categories of modules, and here we shall be concerned with functors with an additional property. Given a functor $F : \mathscr{A} \to \mathscr{B}$ between categories of modules, F is said to be **additive**, if for any morphisms α, α' in $\mathscr{A}$ between the same modules, so that $\alpha + \alpha'$ is defined, we have

$$(\alpha + \alpha')^F = \alpha^F + \alpha'^F.$$

This can also be expressed by saying that the mapping $\mathscr{A}(X,Y) \to \mathscr{B}(X^F, Y^F)$ is a group homomorphism. For example, the hom functors

$h^A : X \mapsto \mathrm{Hom}_R(A, X)$ and $h_A : X \mapsto \mathrm{Hom}_R(X, A)$ are additive functors, as is the tensor product functor $X \mapsto X \otimes A$. On the other hand, the functor $X \mapsto X \otimes X$ is not additive.

Among functors on module categories we shall be particularly interested in those that preserve exact sequences. A functor F will be called **exact** if it transforms any exact sequence

$$A \xrightarrow{f} B \xrightarrow{g} C \tag{4.33}$$

into an exact sequence

$$A^F \xrightarrow{f^F} B^F \xrightarrow{g^F} C^F. \tag{4.34}$$

In words this demands that im f = ker g implies im f^F = ker g^F. The sequence (4.33) is called a **null sequence** if $fg = 0$, i.e. im $f \subseteq$ ker g, and it is clear that any additive functor transforms a null sequence into a null sequence.

The following lemma is useful in checking exactness.

Lemma 4.15

Let a family of modules and homomorphisms be given:

$$A_\lambda \to B_\lambda \to C_\lambda, \quad \lambda \in I; \tag{4.35}$$

write $A = \prod A_\lambda, B = \prod B_\lambda, C = \prod C_\lambda$, and consider the induced sequence

$$A \to B \to C. \tag{4.36}$$

Then (4.36) is exact if and only if all the sequences (4.35) are exact. Similarly for the direct sums.

Proof

Let the mappings in (4.35) be f_λ, g_λ and in (4.36), f, g. Given any $y \in B$, let $y = (y_\lambda)$; it is clear that "$yg = 0$" $\Leftrightarrow$ "$y_\lambda g_\lambda = 0$ for all $\lambda \in I$" and "$y = xf$ for some $x = (x_\lambda) \in A$" $\Leftrightarrow$ "$y_\lambda = x_\lambda f_\lambda$ for some $x_\lambda \in A_\lambda$, for all $\lambda \in I$". It follows that (4.36) is exact iff (4.35) is exact for all $\lambda \in I$. The same proof applies for direct sums. ∎

Exactness is a rather special property of functors, and one usually has to be satisfied with less. Given a short exact sequence

$$0 \to B \xrightarrow{f} A \xrightarrow{g} C \to 0, \tag{4.37}$$

this is transformed by any additive functor F into a null sequence

$$0 \to B^F \xrightarrow{f^F} A^F \xrightarrow{g^F} C^F \to 0. \tag{4.38}$$

The functor F is called **left exact** if (4.38) is always exact except possibly at C^F; if (4.38) is always exact except possibly at B^F, then F is said to be **right exact.** A contravariant functor $G : \mathscr{A} \to \mathscr{B}$ is called **left exact** if the corresponding covariant functor op.$G : \mathscr{A}^\circ \to \mathscr{B}$ is left exact; similarly a **right exact** contravariant functor G is defined by the right exactness of op.G.

In order to check that F is left exact it is enough to show that it preserves kernels, for when (4.37) is given, then $B \cong \ker g$, and (4.38) will be exact at B^F and A^F (and hence F left exact) precisely when $B^F \cong \ker g^F$. Similarly, since exactness of (4.37) at A and C means that $C \cong \operatorname{coker} g$, we can verify that F is right exact iff it preserves cokernels. For this reason a left exact functor is also called "kernel-preserving" and a right exact functor "cokernel-preserving".

The most important example of a left exact functor is the hom functor:

Theorem 4.16

Let R be any ring and ${}_R\mathscr{M}$ its category of left modules. Then the functor Hom_R(-, -) is left exact in each argument.

Proof

Given $g : A \to C$, the kernel of the induced mapping $g^{h^M} : A^{h^M} \to C^{h^M}$ is the set of all homomorphisms $\varphi : M \to A$ annihilated by g, i.e. the mappings whose image lies in $\ker g = A$. This means that $\ker(g^{h^M}) = (\ker g)^{h^M}$, as we had to show. In the same way we have, for any $f : B \to A$,

$$\ker(f^{h_N}) = (\operatorname{coker} f)^{h_N},$$

and this shows the contravariant hom functor to be left exact. ■

Now projective modules can be characterized as follows.

Theorem 4.17

Let R be any ring and P any R-module. Then the following conditions on P are equivalent:

(a) every short exact sequence with P in the third position

$$0 \to A \to B \to P \to 0 \tag{4.39}$$

is split exact;

(b) P is a direct summand of a free module;

(c) every homomorphism from P to a quotient of a module B can be lifted to B; thus the diagram below can be completed by a mapping $f' : P \to B$ to form a commutative triangle:

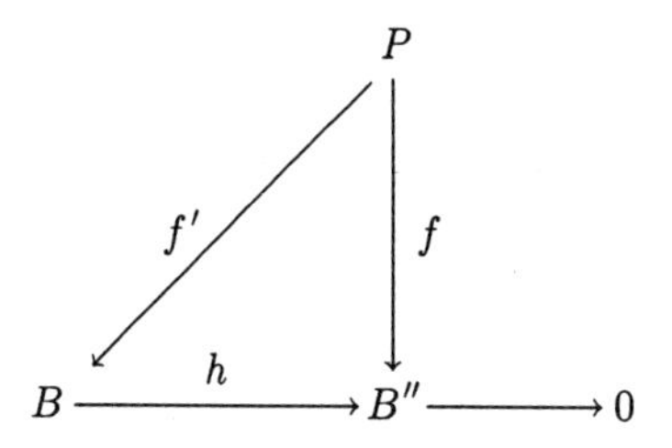

(d) Hom(P, -) is an exact functor.

Proof

(a) $\Rightarrow$ (b). We can write P as a quotient of a free module,

$$0 \to G \to F \to P \to 0; \tag{4.40}$$

by (a) this sequence is split exact, so P is a direct summand of the free module F.

(b) $\Rightarrow$ (c). By (b) we can write $F = P \oplus P'$, where F is free. Given the diagram shown, we obtain a homomorphism $g : F \to B''$ by combining the projection $F \to P$ with the map $f : P \to B''$. We now take a basis $u_1, \ldots, u_r$ of F and choose $b_i \in B$ such that $b_i h = u_i g$. Since h is surjective, this is always possible. Further, because F is free, there is a homomorphism $g' : F \to B$ which maps u_i to b_i, hence $u_i g' h = b_i h = u_i g$ and hence $g' h = g$. If we now combine the injection $P \to F$ with g' we obtain a homomorphism $f' : P \to B$ which is such that $f' h = f$, and this is the required mapping.

(c) $\Rightarrow$ (d). Given a short exact sequence (4.37), say, we apply $h^P = \mathrm{Hom}_R(P, \text{-})$ and obtain

$$\mathrm{Hom}_R(P, B) \to \mathrm{Hom}_R(P, A) \to \mathrm{Hom}_R(P, C) \to 0. \tag{4.41}$$

By Theorem 4.16, exactness can fail only at the third place, $\mathrm{Hom}_R(P, C)$, but by (c) every map $P \to C$ lifts to a map $P \to A$, hence (4.41) is also exact at the third place.

(d) $\Rightarrow$ (a). By applying the functor h^P to (4.39) and remembering (d), we obtain the exact sequence

$$0 \to \mathrm{Hom}_R(P, A) \to \mathrm{Hom}_R(P, B) \to \mathrm{Hom}_R(P, P) \to 0. \tag{4.42}$$

We have $1_P \in \mathrm{Hom}_R(P, P)$ and by exactness there exists $h : P \to B$ which combined with the second map in (4.39) gives the identity on P and thus provides a splitting of (4.39), which proves (a). ■

Given a family, possibly infinite, (P_λ) of projective R-modules, whose direct sum is P, we have the exact functors $\mathrm{Hom}(P_\lambda, \text{-})$, and it is clear that their direct product $\prod \mathrm{Hom}(P_\lambda, \text{-})$ is again exact. By (1.24) of Section 1.4 this means that $\mathrm{Hom}(\oplus P_\lambda, \text{-})$ is exact, and so $\oplus P_\lambda$ is projective. Thus we have proved

Corollary 4.18

The direct sum of any family of projective modules is again projective. ■

By contrast, the direct product of an infinite family of projective modules need not be projective; precise conditions on the ring for this to happen have been found by Chase (1960).

To give examples of projective modules, over a field K (even skew) every projective module is free, but the matrix ring $\mathfrak{M}_n(K)$ (where $n > 1$) has projective modules that are not free. The set of matrices whose columns after the first are zero forms a left ideal C, say; generally the matrices with all columns apart from one, say the i-th, equal to zero form a left ideal isomorphic to C; thus we have $C^n \cong R$. This shows that C is projective, but not free. On the other hand, in a PID every submodule of a free module is free (see Theorem 3.21), and it follows that every projective module is free.

Secondly consider a semisimple ring R. In Section 2.3 we saw that R has the form $R = I_1 \oplus \ldots \oplus I_r$, where each I_i is a minimal ideal, hence simple Artinian as a ring and so a full matrix ring over a skew field. Let P_i be a minimal left ideal in I_i; then it follows that every projective R-module is a direct sum of a number of copies of $P_1, P_2, \ldots$ and P_r, and so it is far from being free.

Unlike a free module a projective module may not have a basis, but it always has a "projective coordinate system" with similar properties. We recall from Section 1.5 that in a free left R-module F with basis $\{u_i\}$ every element x can be written as

$$x = \sum a_i u_i, \quad \text{where } a_i \in R, \tag{4.43}$$

where a_i is uniquely determined by x as the value of the i-th projection from F to R. We shall denote this projection mapping by α_i and write (α_i, x) for its value at x; thus we have $(\alpha_i, x) = a_i$ and (4.43) takes on the form

$$x = \sum (\alpha_i, x) u_i. \tag{4.44}$$

Here $\{u_i\}$ is a basis of F and $\{\alpha_i\}$ is a basis of the dual of F, $F^* = \mathrm{Hom}_R(F, R)$, and these bases are said to be **dual** to each other, because they satisfy the relations

$$(\alpha_i, u_j) = \delta_{ij}, \tag{4.45}$$

as follows by replacing x by u_j in (4.44).

We shall now show that such dual bases exist for any projective module, i.e. bases for which (4.44) holds, although they may not satisfy (4.45). In fact this leads to a characterization of projectives. For simplicity we state the result for finitely generated modules, although it holds without this restriction.

Proposition 4.19 (Dual Basis Lemma)

Given any ring R and a left R-module P with finite generating set $u_1, \dots, u_r$, the module P is projective if and only if there exist $\alpha_1, \dots, \alpha_r \in \mathrm{Hom}_R(P, R)$ such that

$$x = \sum (\alpha_i, x) u_i \quad \text{for all } x \in P. \tag{4.46}$$

Proof

We take a presentation for P. Thus let F be free on $v_1, \dots, v_r$ and define a homomorphism $f : F \to P$ by the rule $v_i \mapsto u_i$, giving the short exact sequence

$$0 \to \ker f \to F \to P \to 0. \tag{4.47}$$

If $\alpha_1, \dots, \alpha_r$ exist as in the statement, satisfying (4.46), then the sequence (4.47) is split by the homomorphism $x \mapsto (x_i)$ from P to $R^r \cong F$, where $x_i = (\alpha_i, x)$; hence P, as a direct summand of F, is projective. Conversely, if P is projective, then there exists $g : P \to F$ splitting (4.47), thus $xgf = x$ for all $x \in P$. Define $\pi_i : F \to R$ as the projection on the i-th factor and put $\alpha_i = g\pi_i : P \to R$. Then $xg = \sum (\pi_i, xg) v_i = \sum (\alpha_i, x) v_i$, hence $x = xgf = \sum (\alpha_i, x) u_i$, which shows that (4.47) holds. ■

This result shows more precisely: if P is finitely generated, by r elements, then it is projective iff it is a direct summand of R^r. For example, a cyclic module is projective iff it is a direct summand of R, but a left ideal of R may well be projective without being a direct summand of R, because it need not be principal.

Exercises 4.5

1. Describe all the finitely generated projective modules over $\mathbb{Z}/(m)$. Which of these modules are generators? (An R-module M is a **generator** if R is a homomorphic image of M^n for some $n \geq 1$.)

2. Show that any finitely generated projective module (over any ring) is finitely presented.

3. Let A be a commutative integral domain with field of fractions K, and let I be an ideal in R. Further, assume that K contains elements $b_1, \ldots, b_r$ such that $Ib_i \subseteq A$ for $i = 1, \ldots, r$ and I contains $a_1, \ldots, a_r$ satisfying $a_1b_1 + \ldots + a_rb_r = 1$. Show that (i) I is generated by $a_1, \ldots, a_r$ and (ii) I is projective. (This result shows that every invertible ideal is projective; moreover, every non-zero projective module is invertible (see A.2, Section 9.5 or Exercise 4 of Section 3.6).)

4. Let R be a commutative integral domain. Show that a projective right ideal in R is free iff it is principal.

5. Show that in the ring $\mathbb{R}[\cos\theta, \sin\theta]$ the ideal generated by $\sin\theta$ and $1 - \cos\theta$ is projective but not principal.

6. Let R be any ring and P a projective R-module with an n-element generating set. Show that there exists P' such that $P \oplus P' \cong R^n$, where the projection $R^n \to P$ is given by an idempotent matrix.

7. If P is as in Exercise 6, with idempotent matrix E, show that P is free iff there exists an invertible matrix U such that $U^{-1}EU = I_r \oplus 0$; what is the significance of r?

8. Use Exercise 7 to show that over a local ring every finitely generated projective module is free.

9. Let R be a matrix local ring. Show that there exists an integer k such that for every projective module P, P^k is free.

10. Given two module homomorphisms $f_i : M_i \to N$ $(i = 1, 2)$ with the same target N, show that there is a module P with homomorphisms $g_i : P \to M_i$ such that (i) $g_1f_1 = g_2f_2$ and (ii) given $h_i : T \to M_i$ such that $h_1f_1 = h_2f_2$, there exists $k : T \to P$ such that $h_i = kg_i$(P is called a **pullback** and the resulting square is a **pullback diagram**). Show also that ker $g_2 \cong$ ker f_1, ker $g_1 \cong$ ker f_2

11. Given any module M and short exact sequences $0 \to Q_i \to P_i \to M \to 0$ $(i = 1, 2)$, where each P_i is projective, show that $P_1 \oplus Q_2 \cong P_2 \oplus Q_1$. (Schanuel's lemma. Hint: use Exercise 10.)

4.6 Injective Modules

Among the equivalent ways of defining projective modules (Theorem 4.7) there was a categorical one: P is projective iff the covariant hom functor Hom(P, -) is exact, and this suggests that we look at the dual notion. Let us define an **injective** module as a module I such that the contravariant hom functor Hom(-, I) is exact. Although in a sense dual to projectives, the injective modules show very different behaviour, as we shall see below.

Our first task will be to see how far Theorem 4.17 can be dualized; by a straightforward translation of the proof we obtain

Theorem 4.20

Let R be any ring and I any R-module. Then the following conditions on I are equivalent:

(a) Hom(-, I) is an exact functor, in other words, I is injective;

(b) every homomorphism from a submodule of a module M into I can be extended to a homomorphism $M \to I$;

(c) every short exact sequence with I in the first position

$$0 \to I \to A \to B \to 0$$

is split exact. In other words, if I is a submodule of a module A, it is a direct summand.

The proof is analogous to that of Theorem 4.17; we shall leave the details to the reader and just indicate the equivalence (a) $\Leftrightarrow$ (b). Property (b) is clearly equivalent to the exactness of the sequence

$$\mathrm{Hom}(M, I) \to \mathrm{Hom}(M', I) \to 0, \tag{4.48}$$

which is just the defining property of injective modules, i.e. (a). ■

As in the case of projective modules, we can use (1.23) of Section 1.4 to prove

Corollary 4.21

The direct product of any family of injective modules is injective. ■

Again it is not the case that the direct sum of injectives is always injective, but we shall obtain a criterion below, for this to happen (Proposition 4.26).

There is an important simplification of condition (b) due to **Baer** (Reinhold Baer, 1902–1979). The problem is to extend a homomorphism from a submodule of M into I to a homomorphism $M \to I$; if this can be done at all, it can be done by adjoining one element at a time, so we need only postulate the extendibility to modules with cyclic quotients. Thus we need to show that every homomorphism from a left ideal of R into I can be extended to a homomorphism $R \to I$. If this homomorphism maps $1 \in R$ to $c \in I$, then $x = x.1$ is mapped to $x.c$; in other words, every homomorphism $R \to I$ has the form $x \mapsto xc$ for some $c \in I$. So the result to be proved takes the following form.

Theorem 4.22 (Baer's Criterion)

Let R be any ring. A left R-module I is injective if and only if every homomorphism $L \to I$, where L is a left ideal of R, can be extended to a homomorphism $R \to I$.

Proof

If I is injective, then as we have seen, every homomorphism $L \to I$ can be extended to map R to I, so the condition is necessary. Suppose now that it is satisfied, and let an R-module M with a submodule M' be given with a homomorphism $f : M' \to I$. Consider the set of homomorphisms with domain contained in M and extending f. They may be partially ordered by writing $f_1 \leq f_2$ whenever the domain of f_2 contains that of f_1 and f_1, f_2 agree on the latter. These extensions form an inductive family, as is easily verified; hence by Zorn's lemma there is a maximal one $f^* : M^* \to I$. If $M^* \neq M$, take $a \in M \setminus M^*$ and put $N = M^* + Ra$. The set $L = \{x \in R | xa \in M^*\}$ is easily seen to be a left ideal of R and the mapping $f_0 : L \to I$ defined by $xf_0 = (xa)f^*$ is a homomorphism from L to I. By hypothesis it can be extended to R, so there exists $c \in I$ such that $xf_0 = xc$. We now extend our homomorphism to N by writing

$$(u + xa)f' = uf^* + xc, \quad \text{where } u \in M^*, x \in R. \tag{4.49}$$

Here the expression $u + xa$ for an element of N may not be unique, because $u + xa = 0$ does not imply $u = 0 = x$; however, if $u + xa = 0$, then $x \in L$ by definition of the latter, hence $uf^* + xc = uf^* + (xa)f^* = (u + xa)f^* = 0$. This shows the mapping f' to be well defined by (4.49). It is easily seen to be a homomorphism from N to I, but this contradicts the maximality of M^*. Therefore $M^* = M$ and f has been extended to M, which shows I to be injective. ■

Condition (c) of Theorem 4.20 leads to another useful criterion for injectivity. Given a module M with a submodule M', the extension $M' \subseteq M$ is said to be **essential**, if M' has a non-zero intersection with every non-zero submodule of M.

Proposition 4.23

Given any ring R, an R-module M is injective if and only if it has no proper essential extension.

Proof

Assume that M is injective and suppose that M is a submodule of N. By Theorem 4.20(c), M is a direct summand and so N is not an essential extension, unless $N = M$. Thus M has no proper essential extensions. Conversely, assume that M has no proper essential extensions and let L be any module containing M as a submodule. The family of all submodules of L meeting M in 0 is clearly inductive, and so, by Zorn's lemma, has a maximal member L_0 say. Consider $\bar{L} = L/L_0$; since $M \cap L_0 = 0$ by construction, M maps isomorphically to a submodule $\bar{M}$ of $\bar{L}$, and by the maximality of L_0 it follows that $\bar{L}$ is an essential extension of $\bar{M} \cong M$. Hence it cannot be proper, so $\bar{L} = \bar{M}$, i.e. $M + L_0 = L$ and $M \cap L_0 = 0$; this shows M to be a direct summand in L and it follows that M is injective. ■

In order to gain some insight into the structure of injective modules we shall need another definition. Let R be an integral domain and M a left R-module; M is said to be **divisible** if the equation in x:

$$u = ax, \quad \text{where } u \in M, a \in R^{\times}, \tag{4.50}$$

always has a solution in M. Every injective module is divisible, for the mapping $ra \mapsto ru$ $(r \in R)$ is a homomorphism from Ra to M and if M is injective, it extends to a homomorphism $R \to M$. If $1 \mapsto v$, then $a \mapsto av$ and since $a \mapsto u$, it follows that (4.50) is solved by $x = v$. Our next result gives conditions for the converse to hold.

Proposition 4.24

Every injective module over an integral domain is divisible. Over a principal ideal domain the converse holds: every divisible module is injective.

Proof

The first part has already been established. To prove the converse we have to show that for a divisible module M over a PID R, any homomorphism f of a left ideal L of R into M extends to a homomorphism $R \to M$. Now L is principal, say $L = Ra$; suppose that f maps a to u, so $f : L \to M$ is of the form $f : ra \mapsto ru$ ($r \in R$). By divisibility, (4.50) has a solution $x = v$; thus $u = av$ and so the mapping $f' : r \mapsto rv$ is a homomorphism extending f, because $(ra)f' = rav = ru$. ■

With the help of this result it is easy to determine all injective $\mathbb{Z}$-modules, i.e. the divisible abelian groups. An example of a torsion-free divisible group is $\mathbb{Q}$, the additive group of rational numbers, or more generally, a direct sum of copies of $\mathbb{Q}$. Another divisible group, in multiplicative notation, is the group of all p^n-th roots of unity, for all n; this is a torsion group, usually denoted by $\mathbb{Z}(p^\infty)$. It can be shown that every divisible abelian group is a direct sum of groups of this type (cf. Fuchs, 1958).

Moreover, every abelian group can be embedded in a divisible abelian group; for example, $\mathbb{Z}$ can be embedded in $\mathbb{Q}$. For a proof of the general case let us take any abelian group A and write it as a quotient of a free abelian group: $A \cong F/N$. Here F is a direct sum of copies of $\mathbb{Z}$ and by embedding each $\mathbb{Z}$ in $\mathbb{Q}$ we can embed F in a vector space over $\mathbb{Q}$, say G. Clearly G is divisible as $\mathbb{Z}$-module, hence so is G/N, and it contains $F/N \cong A$ as a submodule, so A has been embedded in a divisible abelian group.

In Section 3.5 we saw that over a PID every module M has a presentation

$$0 \to G \to F \to M \to 0,$$

where F is free and G, as submodule of F, is also free. For a general ring the submodules of a free module need not be free, or even projective. Of course, since every free module is projective, it is true that every module can be written as quotient of a projective module, a fact expressed by saying that the category ${}_R\mathcal{M}$ has enough projectives. The kernel need not be projective, but we can again present it as a quotient of a projective, and by continuing in this fashion, we obtain an exact sequence

$$\ldots \to P_2 \to P_1 \to M \to 0,$$

called a **projective resolution** of M. This resolution terminates if the mapping from P_n to P_{n-1} has a projective kernel, so that we may take $P_{n+2} = 0$. Homological algebra makes a study of modules having a finite projective resolution; the least length of such a projective resolution is called the **homological** or **projective dimension** of M, written $\mathrm{hd}(M)$, and the supremum of the projective dimensions of R-modules is called the **global dimension** of R, written gl.dim.(R). For example, a polynomial ring in n indeterminates over a field has global dimension n; on the other hand $\mathbb{Z}/(4)$ has infinite global dimension (cf. Exercise 4.6).

Dually one may ask whether every module can be embedded in an injective module. This is in fact true for every ring, thus ${}_R\mathcal{M}$ has **enough injectives**:

Theorem 4.25

Given any ring R, any R-module may be embedded in an injective R-module.

Proof

Consider first the case $R = \mathbb{Z}$. We have just seen that every $\mathbb{Z}$-module A can be embedded in a module which is divisible, but this is the same as being injective, by Proposition 4.24. Thus A has been embedded in an injective $\mathbb{Z}$-module.

In the general case there is a natural homomorphism $h : \mathbb{Z} \to R$ given by $n \mapsto n.1$. Given a right R-module M, we may regard it as a $\mathbb{Z}$-module and so embed it in an injective $\mathbb{Z}$-module I. It follows that we have a homomorphism $\mathrm{Hom}_{\mathbb{Z}}(R, M) \to \mathrm{Hom}_{\mathbb{Z}}(R, I)$, which is an embedding, by the left exactness of Hom. The mapping $h : \mathbb{Z} \to R$ gives rise to a homomorphism

$$\mathrm{Hom}_{\mathbb{Z}}(R, M) \to \mathrm{Hom}_{\mathbb{Z}}(\mathbb{Z}, M) \cong M.$$

Here M is embedded as $\mathbb{Z}$-module in $\mathrm{Hom}_{\mathbb{Z}}(R, M)$ by mapping $m \in M$ to the homomorphism $a \mapsto ma$. Thus M is embedded in $\mathrm{Hom}_{\mathbb{Z}}(R, I)$ and this mapping is R-linear, because $(ma)b = m(ab)$ for $a, b \in R$. To complete the proof we note that $\mathrm{Hom}_{\mathbb{Z}}(R, I)$ is R-injective; this follows because for any R-module N, we have

$$\mathrm{Hom}_R(N\ \mathrm{Hom}_{\mathbb{Z}}(R, I)) \cong \mathrm{Hom}_{\mathbb{Z}}(N \otimes R, I) \cong \mathrm{Hom}_{\mathbb{Z}}(N, I),$$

and this is exact, as a functor in N, because I is injective as $\mathbb{Z}$-module. ∎

With the help of this result we can form an injective resolution of any module, by embedding it in an injective, then embedding the cokernel of this mapping, and so on. This leads to a **cohomological** or **injective dimension** for each module M, written $\mathrm{cd}(M)$. The supremum of these dimensions coincides with the global dimension, as defined in terms of projective dimensions (cf. e.g. A.3, Chapter 3).

As we saw in Corollary 4.21, the direct product of any family of injective modules is again injective; when "product" is replaced by "sum", this no longer holds generally, but there is a simple necessary and sufficient condition; here we merely prove the sufficiency (for the general case see A.3, Section 6.6):

Proposition 4.26

Given any ring R, the direct sum of any family of injective left R-modules is injective provided that R is left Noetherian.

Proof

Assume that R is left Noetherian and for any family $\{E_\lambda\}\,(\lambda \in I)$ of injective modules put $E = \oplus_I E_\lambda$. To show that E is injective it is enough, by Theorem 4.22, to show that any homomorphism $f : L \to E$ from a left ideal L of R into E extends to a homomorphism of R into E. Since R is left Noetherian, L is finitely generated, by $u_1, \ldots, u_r$ say and the images $u_1 f, \ldots, u_r f$ in E have only finitely many non-zero components and so lie in a submodule $E' = \oplus_{I'} E_\lambda$ of E, where I' is a finite subset of the index set I. Thus f maps L into E', but the latter is injective, as a finite direct sum of injective modules. Hence f extends to a homomorphism from R to E' and by combining this with the inclusion $E' \subseteq E$ we obtain the desired extension of f. ∎

Exercises 4.6

1. Given two module homomorphisms $f_i : M \to N_i$ $(i = 1, 2)$ with the same source M, show that there is a module S with homomorphisms $g_i : N_i \to S$ such that (i) $f_1 g_1 = f_2 g_2$ and (ii) given $h_i : N_i \to T$ such that $f_1 h_1 = f_2 h_2$, there exists $k : S \to T$ such that $h_i = g_i k$ (S is called the **pushout** and the resulting square is a **pushout diagram**). Show also that $\mathrm{coker}\, f_1 \cong \mathrm{coker}\, g_2$, $\mathrm{coker}\, f_2 \cong \mathrm{coker}\, g_1$.

2. Given any module M and two short exact sequences $0 \to M \to I_i \to J_i \to 0$, where I_i is injective $(i = 1, 2)$, use Exercise 1 to show that $I_1 \oplus J_2 \cong I_2 \oplus J_1$ (the dual of Schanuel's lemma).

3. Define a **cogenerator** as a module U such that for any module M and any $x \in M$, if $x \neq 0$, there exists $f : M \to U$ such that $xf \neq 0$. Show that an injective R-module I is a cogenerator iff every R-module can be embedded in a suitable power of I.

4. Show that $\mathbb{Q}/\mathbb{Z}$ is an injective cogenerator for $\mathbb{Z}$.

5. Given any ring R, let M be an R-module and I an injective module in which M is embedded. Show that there is a maximal essential extension E of M in I and that E is injective. Show also that E is the least injective submodule of I containing M (E is called the **injective hull** of M).

6. Obtain a projective resolution of the ideal (2) in $\mathbb{Z}/(4)$ by mapping $1 \mapsto 2$ and then resolving the kernel. Show that there is no finite resolution and deduce that (2) has infinite homological dimension.

7. Let $R = \mathbb{Z}/(4)$. Show that R is injective and obtain an injective resolution of (2). What is its cohomological dimension?

4.7 Invariant Basis Number and Projective-Free Rings

We have seen in Section 1.4 that for a (non-trivial) commutative ring every free module has a uniquely determined rank. This need no longer hold for general rings, and we shall say that a ring R has the **invariant basis property** or **invariant basis number** (IBN) if every free module has a uniquely determined rank, in other words, if $R^m \cong R^n \Rightarrow m = n$. A ring lacking this property may be called **non-IBN**. For example, let k be a field and V a vector-space of countable dimension over k. Since the union of two countable sets is again countable, it follows that $V \oplus V \cong V$. Writing $R = \mathrm{End}_k(V)$, we have $R = \mathrm{End}_k(V) = \mathrm{Hom}_k(V, V) \cong \mathrm{Hom}_k(V, V \oplus V) \cong \mathrm{Hom}_k(V, V) \oplus \mathrm{Hom}_k(V, V) \cong R \oplus R$. Thus $R^2 \cong R$ and by induction $R^n \cong R$; so all finitely generated R-modules have the same rank, and IBN fails in a spectacular way.

Let R be a non-trivial non-IBN ring and let m be the least positive integer such that R^m does not have a unique rank, and p the least positive integer such that $R^m \cong R^{m+p}$. Such a ring is said to be of **type** (m, p); it can be shown that for any pair of positive integers m, p a ring of type (m, p) exists (cf., e.g. Cohn, 1985, Section 2.11). Here the trivial ring can be included as a ring of type (0,1). In a ring of type (m, p) we have the isomorphism

$$R^m \cong R^n, \quad \text{where } n = m + p. \qquad (4.51)$$

Let the bases of R^m and R^n, written as row vectors, be $\boldsymbol{u}$, $\boldsymbol{v}$ respectively. Then the isomorphism (4.51) is expressed by the equations $\boldsymbol{v} = \boldsymbol{u}A$, $\boldsymbol{u} = \boldsymbol{v}B$, where $A \in {}^mR^n$ and $B \in {}^nR^m$. Since $\boldsymbol{u} = \boldsymbol{v}B = \boldsymbol{u}AB$, we have $AB = I$, and similarly $BA = I$. Conversely, the existence of two such matrices, where $m \neq n$, means that R lacks IBN. The situation is summed up by

Theorem 4.27

Let R be a non-IBN ring of type (m, p). Then there is an $m \times n$ matrix A over R and an $n \times m$ matrix B, where $n = m + p$, such that

$$AB = I, \quad BA = I. \tag{4.52}$$

Conversely, any ring with matrices of these sizes satisfying (4.52) is non-IBN of type (r, s), where $m \geq r$ and $s \equiv 0 \pmod{p}$.

Only the last part still needs proof, an easy verification that may be left to the reader. ■

This result can be summed up (somewhat imprecisely) by saying that IBN fails in a ring iff the ring contains a pair of rectangular matrices that are mutually inverse.

It can be shown that the ring defined by the entries of the matrices A and B with (4.52) as a complete set of defining relations is an integral domain, provided that $m > 1$ (cf. Cohn, 1995, Section 5.7). This provides examples of integral domains that are not embeddable in skew fields.

Let R be any ring and M a left R-module; we can define a new module by the equation

$$M^* = \mathrm{Hom}_R(M, R). \tag{4.53}$$

Since we have the situation $({}_RM, {}_RR_R)$, it is clear that M^* is a right R-module. Similarly, if M is a right R-module, then M^* is a left R-module. In the case when R is a field and M a finite-dimensional vector space, M^* is just the dual of M that we already met in Section 1.2 and, as we saw in Section 1.5, we then have the isomorphism

$$M^{**} \cong M. \tag{4.54}$$

For general rings this need not hold, but it is easily seen that (4.54) still holds when M is a free R-module of finite rank. For when M is a direct sum, $M = M_1 \oplus M_2$, then it is clear that $M^* = \mathrm{Hom}(M_1 \oplus M_2, R) \cong \mathrm{Hom}(M_1, R) \oplus \mathrm{Hom}(M_2, R) = M_1^* \oplus M_2^*$. Further, $R^* = \mathrm{Hom}_R(R, R) \cong R$, therefore $R^{n*} \cong R^n$; in other words, the dual of a free left R-module of rank

n is a free right R-module of rank n. More generally, let P be a finitely generated projective module and choose P' such that $P \oplus P' \cong R^n$. Then by dualizing, we obtain $P^* \oplus P'^* \cong R^n$, hence P^* is again finitely generated projective, and moreover, by repeating the operation, we find that $P^{**} \cong P$. We state the result as

Theorem 4.28

Let R be any ring. Then the correspondence $P \mapsto P^* = \mathrm{Hom}_R(P, R)$ maps finitely generated projectives to finitely generated projectives and further, $P^{**} \cong P$. ■

It can be shown that the correspondence $M \mapsto M^*$ is a duality (with $M^{**} \cong M$) for all finitely generated modules M, for a large class of Artinian rings (the quasi-Frobenius rings), but this does not concern us here.

Generally speaking, rings without IBN form part of the pathology of rings, and may be disregarded in most situations. We have already seen (in Theorem 1.25) that every (non-trivial) commutative ring has IBN. The next result shows that the same is true of a large class of non-commutative rings.

Theorem 4.29

(i) If a ring R has invariant basis number, then so does any subring of R. (ii) Any non-trivial (left or) right Noetherian ring has invariant basis number.

Proof

The proof of (i) is clear: if a subring S of R fails to have IBN, then it has a pair of mutually inverse rectangular matrices, which lie also in R. (ii) Suppose that IBN fails to hold in a ring R. Then for some $n > m \geq 1$, $R^m \cong R^n$. It follows that R^n is isomorphic to a proper homomorphic image of itself (obtained by setting the last $n - m$ coordinates of every element equal to zero). By repetition we obtain a series of homomorphic images, whose kernels form a strictly ascending chain of submodules of R^n, which therefore cannot be Noetherian. But R^n is finitely generated, hence R cannot be right Noetherian, and by symmetry, neither can it be left Noetherian. ■

We remark that by Corollary 2.28, this result also holds for Artinian rings.

As was mentioned in Section 4.6, homological algebra classifies modules according to their projective resolution, but this still leaves the structure of the projective modules to be elucidated. The simplest case is that where every projective module is free. To avoid pathological situations, we shall also assume that R has invariant basis number. A ring R will be called **projective-free**, if every finitely generated projective module is free, of unique rank. As Theorem 4.28 shows, it does not matter whether we consider left or right projective modules. For example the ring $\mathbb{Z}$ of integers is projective-free, more generally this holds for any principal ideal domain, as we saw in Section 3.4.

Often a slightly wider definition is needed. A ring R is called **projective-trivial**, if there is a projective R-module P such that every finitely generated projective R-module has the form P^n, for a unique integer n. Again it does not matter whether we take the class of left or of right projective modules. Moreover, we have the following reduction to projective-free rings.

Theorem 4.30

A ring R is projective-trivial if and only if it has the form

$$R \cong \mathfrak{M}_n(S), \tag{4.55}$$

for some projective-free ring S and some positive integer n.

Proof

Suppose that R is projective-trivial and let P be the minimal projective left R-module. Since ${}_RR$ is projective, we have

$$R \cong P^n, \tag{4.56}$$

for some positive integer n, unique by hypothesis. Let us write $S = \mathrm{End}_R(P)$; then by taking endomorphism rings in (4.56), we obtain $R \cong S_n$, which is just (4.55). Here S is again projective-trivial and in the isomorphism (4.55), the minimal projective R-module P corresponds to the minimal projective S of S_n, which shows S to be projective-free. Conversely, if R has the form (4.55), where S is projective-free, then a minimal projective S_n-module consists of a single column C in S_n, hence $R \cong C^n$; more generally, any finitely generated projective R-module is a direct sum of copies of C and so R is projective-trivial. ∎

Exercises 4.7

1. Let $R \to S$ be a ring-homomorphism, where S has IBN; show that R also has IBN.

2. Carry out the verification for Theorem 4.27.

3. For any non-IBN ring R show that an integer h exists such that every finitely generated R-module can be generated by h elements. Find a bound for h when R is of type (m, p).

4. Show that for any ring R there is no bound on the number of generators needed for a finitely generated module iff whenever an $m \times n$ matrix A and an $n \times m$ matrix B exist, satisfying $AB = I$, then $m \leq n$. Deduce that such a ring has IBN. (Such a ring is said to have **unbounded generating number, UGN** for short.)

5. Let R be a non-IBN ring and let A, A' be $m \times n$ matrices with two-sided inverses B, B' respectively. Find invertible matrices P, Q such that $A' = PA, B' = QB$.

6. Let R be a non-trivial ring generated by $a_1, \ldots, a_n, b_1, \ldots, b_n$ $(n > 1)$, where $\sum a_i b_i = 1,\ b_i a_j = \delta_{ij}$. Show, by expressing any element of R as a polynomial in the a's and b's, that R is simple.

7. Let R be a non-IBN ring. Show that there exist $m, n \in \mathbb{N}$, $m \neq n$, such that $R_m \cong R_n$.

5
General Rings

The topics introduced in Chapters 1–4 can be pursued further in various directions. We shall not do so, but instead describe in this chapter some aspects of the general theory which will help the reader to understand the basic concepts. Their inclusion is justified by the fact that they are usually only found in specialist accounts but do not require extensive background knowledge.

Since we shall frequently be dealing with skew fields, we shall use the term "field" to mean a division ring which may or may not be commutative, and only sometimes use the term "skew field" for emphasis.

5.1 Rings of Fractions

In Section 1.2 we saw that a commutative ring can be embedded in a field iff it is an integral domain. In the general case the situation is very different; the absence of zero-divisors is necessary but, as we saw in Section 4.7, not sufficient for a field of fractions to exist. Secondly, even when a ring R can be embedded in a skew field, the least such field is by no means always unique up to isomorphism, and moreover, its elements cannot always be written in the form ab^{-1}, where $a, b \in R$. There are different ways of dealing with these problems; in the present section we shall concentrate on the last point, by giving conditions for the elements ab^{-1} to form a field.

Given any ring R, by a **field of fractions** of R we understand a field K with an embedding $R \to K$ such that K is generated, as a field, by the image of R. It is clear that for a field of fractions of R to exist, R must be an integral domain. We ask further: when is there a field of fractions for R whose elements are all of the form ab^{-1}, $a, b \in R, b \neq 0$? If such a field is to exist, every element of the form $b^{-1}a$ must be expressible in the form $a'b'^{-1}$; we may assume that $a \neq 0$, since the condition is vacuous for $a = 0$. Multiplying up, we thus obtain the following condition: given $a, b \in R^\times$, there exist a', $b' \in R$, such that $ab' = ba' \neq 0$, or more concisely,

$$\text{O.1 For any } a, b \in R^\times,\ aR \cap bR \neq 0. \tag{5.1}$$

This condition turns out to be sufficient for a field of fractions to exist, and to be unique up to isomorphism. This result was proved by Ore in 1931 and O.1 is known as the **right Ore condition** (after Oystein Ore, 1899–1968) and it constitutes our first result:

Theorem 5.1

Let R be any ring. Then R has a skew field of fractions K such that every element of K can be expressed as ab^{-1}, where $a, b \in R, b \neq 0$, if and only if R is an integral domain satisfying the right Ore condition (5.1). Moreover, K is then determined up to a unique isomorphism over R.

Proof

We have seen that all the conditions are necessary. Suppose that they are satisfied and consider the set $S = R \times R^\times$; we define an equivalence on S by the rule

$$(a, b) \sim (a', b') \Leftrightarrow \text{ for some } u, u' \in R, au = a'u' \text{ and } bu = b'u' \neq 0. \tag{5.2}$$

To show that this is an equivalence we have to verify that it is reflexive, symmetric and transitive. We see that it is reflexive by taking $u = u' = 1$, and symmetry follows from the symmetry of the condition (5.2). To prove transitivity, if $(a, b) \sim (a', b'), (a', b') \sim (a'', b'')$, then $au = a'u', bu = b'u' \neq 0$ and $a'v = a''v', b'v = b''v' \neq 0$. By (5.1) there exist $s, s' \in R$ such that $s' \neq 0$ and $b'u's = b'vs'$ and since R is an integral domain, $b'vs' \neq 0$; moreover, we can cancel b' and find that $u's = vs'$. Now we have $aus = a'u's = a'vs' = a''v's'$ and $bus = b'u's = b'vs' = b''v's' \neq 0$, and it follows that $(a, b) \sim (a'', b'')$, so that (5.2) defines an equivalence. Let us denote the equivalence class of (a, b) by $[a/b]$ and define a ring structure on the set of these classes by the equations

$$[a/b] + [a'/b'] = [ac + a'c'/bc],$$
$$\text{where } c \in R, c' \in R^{\times} \text{ are such that } bc = b'c', \qquad (5.3)$$
$$[a/b].[a'/b'] = [ad'/b'd],$$
$$\text{where } d \in R, d' \in R^{\times} \text{ are such that } a'd = bd'. \qquad (5.4)$$

Of course we must verify that these definitions depend only on the equivalence classes and not on the representatives chosen, but this is easily done if we bear in mind that $[a/b] = [ac/bc]$ for any $c \in R^{\times}$, as follows by taking $u = c, u' = 1$ in (5.2). The verifications are tedious but straightforward, and so will be left to the reader, who should at least carry them out once; they are slightly eased by noting that the definitions (5.3), (5.4) in a special case become (i) $[a/b] + [a'/b] = [a + a'/b]$ and (ii) $[a/b][b/b'] = [a/b']$. In this form it is easier to verify their independence of the choice of representatives and the condition (5.2) can always be used to bring them to this form.

It remains to verify the ring laws, and this will again be left to the reader. The ring so obtained will be denoted by K. It is in fact a field, for it is easily checked that $[a/b] \neq 0$ iff $a \neq 0$, and then $[a/b]^{-1} = [b/a]$. Now it is clear that the mapping $f : a \mapsto [a/1]$ defines an embedding of R in K and if $g : R \to L$ is any embedding of R in a field L, then the elements $(ag)(bg)^{-1}$ for $a \in R$, $b \in R^{\times}$ form a subfield L' of L and the correspondence $af \leftrightarrow ag$ clearly extends to an isomorphism between K and L' over R. ■

An integral domain satisfying the right Ore condition (5.1) is called a **right Ore domain**; left Ore domains are defined similarly and a domain satisfying both the left and right Ore conditions is an **Ore domain**. Any element of the field of fractions can be written as ab^{-1}, or simply a/b, where a will be called the **numerator** and b the **denominator**. Of course we must bear in mind that these terms refer to the expression a/b and not to the element it represents (which may be expressible in other ways). We note that by Theorem 4.29 every Ore domain has IBN.

Theorem 5.1 has an interesting generalization obtained by replacing $R^{\times}$ by a subset S of $R^{\times}$. This set S will have to be **multiplicative**, i.e. it should contain 1 and be closed under multiplication. The Ore condition now takes the form:

O.1′ Given $a \in R, b \in S$, we have $aS \cap bR \neq \emptyset$.

Now it is not necessary to assume that R is an integral domain, but merely

O.2 All elements of S are left and right regular.

For these subsets we have the following result.

Theorem 5.2

Let R be any ring and S a multiplicative subset of R satisfying O.1′ and O.2. Then R can be embedded in a ring T in which all elements of S have inverses and if $f : R \to T$ is the embedding, then every element of T has the form $(af)(bf)^{-1}$, where $a \in R$, $b \in S$.

The proof is exactly along the lines of the proof of Theorem 5.1 and so will not be repeated. ∎

A set satisfying O.1′ and O.2 will be called a **regular right Ore set**; the ring constructed in Theorem 5.2 will be denoted by R_S and is called the **localization** of R at S. In R_S we can operate very much as in a field, bearing in mind that not all elements are invertible. For example, we have the following property, familiar from the commutative case.

Proposition 5.3

Let R be any ring, S a regular right Ore set in R and R_S the localization. In particular, when R is a right Ore domain, S may be taken to be $R^{\times}$. Then any finite set in R_S can be brought to a common denominator.

Proof

Given a/b, $a'/b' \in R_S$, we have seen in the proof of Theorem 5.1 that they can be brought to a common denominator, say $bu' = b'u = m$ and then $a/b = au'/m$, $a'/b' = a'u/m$. In the general case we use induction on the number of elements. Thus if $a_1/b_1, \ldots, a_n/b_n$ are given, we can by induction bring $a_2/b_2, \ldots, a_n/b_n$ to a common denominator, say $a_2'/m, \ldots, a_n'/m$. By the Ore condition we can find $u \in R, u' \in S$ such that $b_1 u = mu'$ and here $mu' \in S$. Hence $a_1/b_1 = a_1 u/b_1 u = a_1 u/mu'$, $a_i/b_i = a_i'/m = a_i'u'/mu'$ for $i = 2, \ldots, n$ and so all n elements have been brought to a common denominator. ∎

The above results show that for Ore domains the theory of fields of fractions is quite similar to that for commutative integral domains; it only remains to find some examples. Later we shall see that they actually form quite a special class of rings, but in Section 5.2 we shall describe a fairly wide class of Ore domains. For the moment we shall establish an important sufficient condition due to **Goldie** (Alfred Goldie, 1920–):

Theorem 5.4

Any integral domain which is right Noetherian is a right Ore domain.

Proof

Let R be an integral domain and assume that R does not satisfy the right Ore condition. Then there exist $a, b \in R^{\times}$ such that $aR \cap bR = 0$. We shall show that R is not right Noetherian by verifying that the chain

$$bR \subset bR + abR \subset bR + abR + a^2bR \subset \ldots$$

is strictly ascending and does not break off. For otherwise we would have $a^{n+1}bR \subseteq bR + abR + \ldots + a^nbR$, say

$$a^{n+1}b = \sum_{i=0}^{n} a^i bc_i.$$

If $a^r bc_r$ is the first non-zero term on the right, we can cancel a^r and so find that

$$a^{n+1-r}b = bc_r + \ldots + a^{n-r}bc_n,$$

but this contradicts the fact that $aR \cap bR = 0$, and the result follows. ∎

We recall from Section 3.5 that a module M over a commutative integral domain is **torsion-free** if each non-zero element generates a free submodule and it is a **torsion module** if every element is a torsion element, but for modules over a general ring the torsion elements need not even form a submodule. However, when R is a right Ore domain, then in any right R-module M the set $\mathrm{t}M$ of all torsion elements is indeed a submodule. For if $x \in \mathrm{t}M$, suppose that $xa = 0$ for $a \in R^{\times}$, and let $b \in R^{\times}$. Then there exist $a', b' \in R$ such that $ab' = ba' \neq 0$; hence $xba' = xab' = 0$, which shows that $xb \in \mathrm{t}M$ for all $b \in R^{\times}$. To show that $\mathrm{t}M$ is closed under addition, let $x, y \in \mathrm{t}M$, say $xa = yb = 0$ and again take a', b' such that $c = ab' = ba' \neq 0$. Then $(x + y)c = xab' + yba' = 0$, hence $x + y \in \mathrm{t}M$.

Let K be the field of fractions of the right Ore domain R and let M be any right R-module. Then we can form the tensor product $M \otimes_R K$, but naturally the torsion elements in M will then be annihilated. Our next result shows that nothing more is annihilated in this process, and it incidentally gives another proof that the set $\mathrm{t}M$ of all torsion elements is a submodule:

Proposition 5.5

Let R be a right Ore domain and K its field of fractions. Then for any right R-module M, the K-space $M \otimes_R K$ may be described as the set of formal products $xb^{-1}(x \in M, b \in R^{\times})$ subject to the relations $xb^{-1} = x'b'^{-1}$ if and only if $xu' = x'u$, $bu' = b'u$ for some $u, u' \in R^{\times}$. The kernel of the canonical mapping

$$M \to M \otimes K \tag{5.5}$$

is the torsion submodule $\mathrm{t}M$, hence (5.5) is an embedding if and only if M is torsion-free.

Proof

Any element of $M \otimes K$ has the form $x = \sum x_i \otimes a_i b_i^{-1}$, where $x_i \in M$, $a_i, b_i \in R^{\times}$. By Proposition 5.3 all these fractions can be brought to a common denominator, say $b_i c_i = b$ and so

$$x = \sum x_i \otimes a_i c_i b^{-1} = \left(\sum x_i a_i c_i\right) b^{-1}.$$

This shows every element of $M \otimes K$ to be of the form yb^{-1}, where $y \in M, b \in R^{\times}$. Further, if $p = yb^{-1}$ and $p' = y'b'^{-1}$, then $bu' = b'u = c$ for some $u, u' \in R^{\times}$, and so we have $pc = yu', p'c = y'u$. From this it is clear that $p = p'$ iff $pc = p'c$, i.e. $yu' = y'u$. It follows that $yb^{-1} = 0$ iff $xu = 0$ for some $u \in R^{\times}$, i.e. precisely when $y \in \mathrm{t}M$. Hence the kernel of (5.5) is $\mathrm{t}M$, which shows this set to be a submodule, and (5.5) is an embedding iff $\mathrm{t}M = 0$, i.e. M is torsion-free. ■

For any right R-module M over a right Ore domain R we have a notion of rank, defined as follows. By Proposition 5.5 we can form $M \otimes K$, which as K-module has a well-defined dimension $(M \otimes K : K)$. We now define the rank of M as

$$\mathrm{rk}\, M = (M \otimes_R K : K). \tag{5.6}$$

We emphasize that this rank has been defined only for modules over Ore domains. From Proposition 5.5 it is clear that $\mathrm{rk}\ M = 0$ iff M is a torsion module.

Exercises 5.1

1. Formulate an Ore condition for monoids and prove an analogue of Theorem 5.1.

2. Let R be a right Ore domain and K its field of fractions. Verify that K is an R-module and show that every finitely generated submodule of K qua right R-module, is free.

3. Let R be a right Ore domain. Given a short exact sequence $0 \to M' \to M \to M'' \to 0$ of right R-modules, show that $\text{rk } M = \text{rk } M' + \text{rk } M''$. What is the generalization to longer exact sequences?

4. Let R be a right Ore domain with field of fractions K. Show that an endomorphism α of R extends to an endomorphism of K iff α is injective, and that the extension is unique.

5. Let R be an integral domain. Show that any right Ore subring is contained in a maximal right Ore subring of R.

6. An integral domain in which every finitely generated right ideal is principal is called a **right Bezout domain**. Verify that every right Bezout domain is right Ore.

7. Let R be a right Ore domain and K its field of fractions. Show that the centre of K consists of all $ab^{-1} (b \neq 0)$ such that $axb = bxa$ for all $x \in R$.

8. An element c in an integral domain R is called **right large** if for every non-zero right ideal A, $cR \cap A \neq 0$. Show that the set of right large elements in R is multiplicative.

9. Let R be a ring with IBN and S a right Ore set in R. Show that the localization R_S need not have IBN.

10. Let K be a skew field and F a subfield of its centre. Given a commutative field extension E/F, if $K \otimes_F E$ is an integral domain, what are the conditions for it to be a right Ore domain?

5.2 Skew Polynomial Rings

In Section 3.1 we defined polynomial rings. For a given commutative ring R the polynomial ring $R[x]$ may be described as the "free extension" of R by a single element, where "free" means that no relations are imposed, beyond those needed to ensure that we get a commutative ring. When we give up commutativity, many problems arise, which are best dealt with one at a time. We saw in Section 5.1 that to form the field of fractions, matters can be simplified if we demand that all elements can be expressed as fractions ab^{-1}, as in the commutative case. In the same way we now consider extensions of the general ring R by an element x, such that the elements in the extension take the form

$$f = x^n a_0 + x^{n-1} a_1 + \ldots + a_n, \quad \text{where } a_i \in R. \tag{5.7}$$

Such an expression will again be called a **polynomial**, and if $a_0 \neq 0$, its **degree**, denoted by $\deg f$, is defined as n. The addition of these polynomials is termwise, as in the commutative case, but when it comes to multiplication, we need to find an expression in the form (5.7) for bx, where $b \in R$. Once this is done, we can carry out general multiplications by making use of the associative and distributive laws. As a further simplification we shall assume that the degree is preserved; thus we have to find an expression of degree 1 for bx. In its most general form this will be

$$bx = xb_0 + b_1, \quad \text{where } b_0, b_1 \in R, b_0 \neq 0 \text{ whenever } b \neq 0. \tag{5.8}$$

It is clear that b_0, b_1 are uniquely determined by b, because the expression on the right is unique; so we shall denote the mappings $b \mapsto b_0$ by α and $b \mapsto b_1$ by δ. On equating the expressions for $bx + cx$ and $(b + c)x$, we find $x(b\alpha) + b\delta + x(c\alpha) + c\delta = x(b + c)\alpha + (b + c)\delta$, hence on equating coefficients of x and of 1, we obtain

$$(b + c)\alpha = b\alpha + c\alpha, \quad (b + c)\delta = b\delta + c\delta \quad \text{for all } b, c \in R. \tag{5.9}$$

If R is a k-algebra, we would expect α and δ to commute with multiplication by scalars, so that α and δ are linear mappings; this amounts to assuming that on k, α reduces to 1 and δ to 0. We also need to equate the expressions for $(bc)x$ and $b(cx)$, and so find that $x(bc)\alpha + (bc)\delta = x(b\alpha)(c\alpha) + b(c\delta) + (b\delta)(c\alpha)$; hence

$$(bc)\alpha = (b\alpha)(c\alpha) \quad \text{for all } b, c \in R \tag{5.10}$$

and

$$(bc)\delta = b(c\delta) + (b\delta)(c\alpha) \quad \text{for all } b, c \in R. \tag{5.11}$$

From (5.9) and (5.10) we see that α is an endomorphism of R, which must be injective, because the degree was preserved in (5.8). As for δ, if we had $\alpha = 1$, then (5.11) would reduce to $(bc)\delta = b(c\delta) + (b\delta)c$; this equation, with linearity, is usually expressed by saying that δ is a **derivation**. A linear mapping δ on R satisfying (5.11) is called an α-**derivation**. These results can be summed up as follows:

Theorem 5.6

Let R be any ring. Given any endomorphism α of R and any α-derivation δ, we can define a ring structure on the set of all polynomials (5.7) over R by means of the commutation rule

$$bx = x(b\alpha) + b\delta \quad \text{for all } b \in R, \tag{5.12}$$

and the resulting ring is an extension of R. Conversely, the set of polynomials with this commutation rule forms a ring where α is an endomorphism and δ is an α-derivation.

Here the last part is to be interpreted as follows: For any mappings α, δ of R into itself the rule (5.12) together with linearity can be used to reduce every product of polynomials to a single polynomial, but the structure so obtained will be a ring extension of R iff α is an endomorphism and δ is an α-derivation. The necessity of these conditions has been proved above; their sufficiency is a simple verification, which will be left to the reader. We remark that it is not necessary to assume α injective to obtain a ring extension, but now the term bx may no longer have degree 1, namely when $b\alpha = 0$. Strictly speaking, it also needs to be shown that in this construction all the powers of x are linearly independent over R, but this follows as in the commutative case by using the representation as infinite sequences of terms, almost all zero. ∎

The ring so obtained is called a **skew polynomial ring** in x over R and is denoted by $R[x; \alpha, \delta]$. Important special cases are obtained by taking $\delta = 0$ or $\alpha = 1$. If both these conditions hold, we just have the usual polynomial ring $R[x]$; the ring $R[x; 1, \delta]$ is sometimes called a **differential polynomial ring**, the ring $R[x; \alpha, 0]$, often simply written $R[x; \alpha]$, is called a **twisted polynomial ring**. Each skew polynomial has a degree as defined above, but we need to impose some restrictions to ensure that the degree has the familiar properties. By convention we shall take the degree of 0 to be $-\infty$; then the required properties are as follows:

Proposition 5.7

Let R be an integral domain with an injective endomorphism α and an α-derivation δ. Then there is a skew polynomial ring $R[x; \alpha, \delta]$ with a degree function deg satisfying the conditions

(i) $\deg(f) \in \mathbb{N}$ for $f \neq 0$, while $\deg(0) = -\infty$;

(ii) $\deg(f - g) \leq \max\{\deg(f), \deg(g)\}$;

(iii) $\deg(fg) = \deg(f) + \deg(g)$.

Moreover, $R[x; \alpha, \delta]$ is an integral domain.

Proof

It is clear that (i) and (ii) hold. To prove (iii), we observe that the skew polynomial ring is defined by the commutation rule (5.12), and the degree is preserved by this relation, because α is injective. It follows that when we form a product fg, where $\deg(f) = r$, $\deg(g) = s$, the product fg has a unique leading term, which is of degree $r + s$, and this remains unchanged under the reduction to normal form, so (iii) follows. Now the last assertion is an easy consequence of (iii). ■

When our ground ring is a skew field K, any endomorphism α is necessarily injective, because ker α is an ideal not containing 1. Thus the conditions of Proposition 5.7 are automatically satisfied; but we can say even more in this case:

Theorem 5.8

Let K be any skew field with an endomorphism α and an α-derivation δ. Then the skew polynomial ring $K[x; \alpha, \delta]$ is a principal right ideal domain, but it is not left principal unless α is an automorphism.

Proof.

Write $P = K[x; \alpha, \delta]$ and note that P is an integral domain by Proposition 5.7 and the above remark. To prove P right principal we shall use the division algorithm, as in the commutative case. Thus let $f, g \in P$, where $g \neq 0$. We claim that $q, r \in P$ exist such that

$$f = gq + r, \qquad \deg(r) < \deg(g). \tag{5.13}$$

This is clear if $\deg(f) < \deg(g)$, so assume that $\deg(g) \leq \deg(f)$, say $f = x^r a_0 + \ldots, g = x^s b_0 + \ldots$; then $r \geq s$ and if $q_1 = b_0^{-1} x^{r-s} a_0$, then gq_1 has the same leading term as f and so $\deg(f - gq_1) < \deg(f)$. If $\deg(f - gq_1) \geq \deg(g)$, we can continue the process and after a finite number of steps (at most $r - s$), we obtain a remainder of degree less than $\deg(g)$ and (5.13) follows. Now this algorithm can be used as in the commutative case to show that every right ideal is principal.

When α is an automorphism, with inverse β, then we have the formula opposite to (5.12):

$$xa = (a\beta)x - a\beta\delta, \tag{5.14}$$

which is obtained from (5.12) by replacing b by $a\beta$ and rearranging the terms. So in this case P is also left principal. If α is not an automorphism, we can find

c in K which is not in the image of α. We claim that

$$Px \cap Pxc = 0, \tag{5.15}$$

so that P is not even a left Ore domain. For suppose that $fx = gxc$, where $f = \sum x^i a_i$, $g = \sum x^i b_i$. If $\deg(f) = r$, say, then on equating highest terms in (5.15) we obtain $x^{r+1}(a_r\alpha) = x^{r+1}(b_r\alpha)c$, hence $c = (b_r^{-1}a_r)\alpha$, which is a contradiction, so (5.15) holds. This shows that P is not a left Ore domain and a fortiori it is not left principal. ■

Theorem 5.8 shows that any skew polynomial ring $K[x;\alpha,\delta]$ over a field K has itself a field of fractions, which is denoted by $K(x;\alpha,\delta)$ and is called the **skew function field**.

Let us show next that any skew polynomial ring over a right Ore domain is again right Ore; of course we must remember that the situation is not symmetric because in the formation of the skew polynomial ring we chose to put the coefficients on the right.

Proposition 5.9

Let R be a right Ore domain with an injective endomorphism α and an α-derivation δ. Then the skew polynomial ring $R[x;\alpha,\delta]$ is again a right Ore domain.

Proof

From Proposition 5.7 we know that $R[x;\alpha,\delta]$ is an integral domain. Since R is a right Ore domain, it has a field of fractions K. Now we can define α on K by the equation $(ab^{-1})\alpha = (a\alpha)(b\alpha)^{-1}$; here it has to be checked that this definition does not depend on the representation of the element ab^{-1}, but this is straightforward and so may be left to the reader. In order to extend δ to K we apply δ to the equation $bb^{-1} = 1$ and obtain $b(b^{-1})\delta + (b\delta)(b\alpha)^{-1} = 0$, hence

$$(b^{-1})\delta = -b^{-1}(b\delta)(b\alpha)^{-1}. \tag{5.16}$$

Now it is easily checked that the operation defined on K as

$$(ab^{-1})\delta = -ab^{-1}(b\delta)(b\alpha)^{-1} + (a\delta)(b\alpha)^{-1} \tag{5.17}$$

is independent of the choice of representative and is an α-derivation. We can thus form the skew polynomial ring $K[x;\alpha,\delta]$. By Theorem 5.8 this is a principal right ideal domain and, as we saw, it has the skew rational function field $K(x;\alpha,\delta)$ as field of fractions. Any skew rational function u has the form fg^{-1}, where f, g are skew polynomials over K. Their coefficients are thus rational functions over R, which by Proposition 5.3 can all be brought to a common denominator. If this denominator is c, this means that f, g have the forms $f = f_1c^{-1}, g = g_1c^{-1}$, where f_1, g_1 are skew polynomials over R. Hence we obtain

$$u = fg^{-1} = f_1c^{-1}cg_1^{-1} = f_1g_1^{-1}.$$

Thus the right quotients of skew polynomials form a skew field and so by Theorem 5.1, $R[x; \alpha, \delta]$ is a right Ore domain, and we have shown incidentally that its field of fractions is also $K(x; \alpha, \delta)$. ■

Let us give some examples of skew polynomial rings.

1. The field $\mathbb{C}$ of complex numbers has an automorphism of order two, complex conjugation $\alpha : a + b\mathrm{i} \mapsto a - b\mathrm{i}$. The twisted polynomial ring $P = \mathbb{C}[x; \alpha]$ consists of all polynomials with complex coefficients and commutation rule $ax = x(a\alpha)$. The centre of this ring is $\mathbb{R}[x^2]$, the ring of all real polynomials in x^2. It is easily verified that $x^2 + 1$ is irreducible in P (although reducible as an ordinary polynomial over $\mathbb{C}$). The residue class ring $P/(x^2 + 1)$ is an algebra over $\mathbb{R}$, spanned by 1, i, and the cosets of x, $x\mathrm{i}$. Denoting the latter by k, j, respectively, we find that $\mathrm{ij} = -\mathrm{ji} = \mathrm{k}, \mathrm{jk} = -\mathrm{kj} = \mathrm{i}$, $\mathrm{ki} = -\mathrm{ik} = \mathrm{j}$. Thus $P/(x^2 + 1)$ is just $\mathbb{H}$, the skew field of quaternions which we already met in Section 2.1. More generally, let E be a field with an automorphism α of order n and denote the fixed subfield by F. Then E/F is a field extension of degree n and in characteristic zero (or at least prime to n) E is generated by an element u satisfying an equation $x^n - a = 0$ over F (cf. A.2, Chapter 3). We take the twisted polynomial ring $P = E[x; \alpha]$ and form the residue class ring $A = P/(x^n - a)$. This is an algebra of dimension n^2 over F; if the coset of x is denoted by u, then A may be described as an n-dimensional vector space over E, with basis (as right E-module), $1, u, \ldots, u^{n-1}$, commutation rule $cu = u(c\alpha)$, and multiplication

$$u^iu^j = \begin{cases} u^{i+j} & \text{if } i + j \leq n, \\ au^{i+j-n} & \text{if } i + j \geq n. \end{cases}$$

The algebra A so obtained is a skew field with centre F and of dimension n^2 over F; it is also called a **division algebra**, more precisely, a **cyclic** division algebra (because a maximal subfield has a cyclic Galois group over F). It can be shown that for every division algebra the dimension over its centre is a square (essentially because over an algebraically closed ground field extension the division algebra becomes a full matrix ring, see A.3, Section 7.1).

2. Let $K = F(t)$ be the field of rational functions in a single variable t; the mapping $\delta : f \mapsto \mathrm{d}f/\mathrm{d}t$ is easily verified to be a derivation of f. In this way we obtain the ring of differential polynomials $D = K[x; 1, \delta]$. We remark that D may be defined as the algebra over F generated by x and t with the defining relation $tx - xt = 1$; it is also known as the **first Weyl algebra** (after Hermann Weyl, 1885–1955).
3. Let K be any field with an automorphism α and consider the twisted polynomial ring $K[x; \alpha]$. The set of powers in x, $\{1, x, x^2, \ldots\}$, is easily seen to be a right Ore set and the localization R may be written $K[x, x^{-1}; \alpha]$. It is called the ring of **twisted Laurent polynomials**. More generally, if K is a field with an automorphism α of finite order n, then for any cyclic group of order m divisible by n, with generator x, we can define the ring generated by x over K with commutation relation $ax = x(a\alpha)$. This suggests a further generalization: Let K be any field on which a group G acts by automorphisms, thus each $\alpha \in G$ defines an automorphism of K, whose action on $c \in K$ will be denoted by c^α, such that $(a^\alpha)^\beta = a^{\alpha\beta}$ for all $a \in K$, $\alpha, \beta \in G$. Then we can form the **skew group algebra** or **crossed product** $K * G$ of G over K, whose underlying space is just the usual group algebra KG (as defined in Section 2.1), with multiplication determined by the commutation rule $a\alpha = \alpha a^\alpha$.

Exercises 5.2

1. Verify that the mapping δ defined on K by (5.17) is independent of the choice of representatives for the fraction and that it is an α-derivation on K.

2. Let $R = K[x; \alpha, \delta]$ be a skew polynomial ring. Show that if $\alpha\delta = \delta\alpha$, then α may be extended to an endomorphism of R by putting $x^\alpha = x$. Show that the subring generated by x^δ over K is $K[x^\delta; \alpha^2]$.

3. Let K be a ring with an injective endomorphism α. Show that K can be embedded in a ring K^* with an automorphism extending α, and that the least such ring is unique up to K-ring isomorphism. If K is a field and $R = K(x; \alpha)$ is the function field on the twisted polynomial ring, show that $K^* = \bigcup x^n K x^{-n}$.

4. Let $q = p^r$, where p is a prime and denote by $\mathbb{F}_q$ the field of q elements, with automorphism $\alpha : c \mapsto c^p$ (it can be shown that such a field exists for every prime power, see A.2, Section 3.8). Define an action of $\mathbb{F}_q$ on $\mathbb{F}_q[x]$ by right multiplication and let T

act by α; show that each polynomial $\sum T^i a_i$ defines an endomorphism of $\mathbb{F}_q[x]$ and verify that the ring of these endomorphisms is the twisted polynomial ring $\mathbb{F}_q[T;\alpha]$.

5. Verify that the crossed product $K * G$ defined in the text is an associative K-ring. If N is a normal subgroup of G with complement L, show that the group ring KG can be expressed as a crossed product $R * L$, where $R = KN$ and L acts on KN by inner automorphisms.

6. Let K be a skew field and $R = K[x, y]$ the polynomial ring in two central indeterminates. Given $a, b \in K$ such that $c = [a, b] = ab - ba \neq 0$, verify that $[x + a, y + b] = c$ and deduce that the right ideal $(x + a)R \cap (y + b)R$ is not free, but becomes free if we form the direct sum with R.

7. Let K be a commutative field with an endomorphism α and an α-derivation δ, and form the skew polynomial ring $R = K[x; \alpha, \delta]$. Find the conditions for R to have an anti-automorphism which fixes K and maps x to $-x$.

8. Let K be a commutative field with an automorphism α. Determine the centre of $K(x; \alpha)$ when α is of infinite order.

9. Let A be a ring with an endomorphism α and an α-derivation δ. Verify that the mapping $\varphi : c \mapsto \begin{pmatrix} c & c^\delta \\ 0 & c^\alpha \end{pmatrix}$ is a homomorphism $A \to A_2$. If A is a right Ore domain with skew field of fractions K and α is injective, show that φ extends to a homomorphism $K \to K_2$.

5.3 Free Algebras and Tensor Rings

The skew polynomial rings in Section 5.2 were obtained by imposing a commutation rule to ensure that polynomials had the same simple form as in the commutative case. An obvious next step is to examine ring extensions when no restrictions are imposed. It turns out to be practical to split the problem and consider first the case of several indeterminates which do not commute with each other but which commute with elements of the ground field. This leads to free algebras, a class of rings whose study has begun only recently and which has revealed some remarkable properties.

Let k be a commutative field and $X = \{x_1, \dots, x_n\}$ a finite set. The polynomial ring $k[x_1, \dots, x_n]$ has been much studied and is well known; we shall not treat it in detail here, but merely remark that its elements can be expressed in the form of finite sums

$$f = \sum_{i_1 \dots i_n} a_{i_1 \dots i_n} x_1^{i_1} \dots x_n^{i_n}. \tag{5.18}$$

Clearly this ring is an integral domain, which is Noetherian, by Corollary 3.4. However, when $n > 1$, its ideals need not be principal, nor even projective. In fact each module over the ring has a projective resolution (cf. Section 4.6) which ends after at most n steps and in some cases n steps are needed. This is expressed by saying that its **global dimension** is n. By contrast we shall find that the free algebras all have global dimension one.

The **free associative algebra** on X over k, denoted by $k\langle X\rangle$, is defined as the k-algebra generated by X with no defining relations. Thus each element has the form

$$f = \sum_{i_1 \dots i_r} a_{i_1 \dots i_r} x_{i_1} \dots x_{i_r}, \tag{5.19}$$

where each suffix i_ρ runs from 1 to n and the sum is over a finite set of suffixes. It is more convenient to abbreviate the suffix set $\{i_1, \dots, i_r\}$ as I and the product of x's in (5.19) as x_I; then the sum may be written simply as $\sum_I a_I x_I$. To each term we shall assign as degree the number of factors x_i, thus the term written in (5.19) has degree r, and the degree of the whole expression is the largest degree of the terms that occur. This degree function again satisfies the conditions (i)–(iii) of Proposition 5.7; the proof is even easier than for Proposition 5.7 (and so will be left to the reader), since there is no commutation rule to worry about. In particular, it follows again that $k\langle X\rangle$ is an integral domain, though of course not commutative, nor even an Ore domain, when $n > 1$. Here the number n of variables is called the **rank** of $k\langle X\rangle$. It is independent of the choice of basis, for on writing $R = k\langle X\rangle$ and denoting the commutator ideal (generated by all commutators $[a, b] = ab - ba$) by C, we have $R/C \cong k[x_1, \dots, x_n]$ and the latter ring determines n as the least number of elements in a generating set.

We note the following property, which has no analogue in the commutative case:

Proposition 5.10

Given a free algebra $k\langle X\rangle$ in a set $X = x_1, \dots, x_n$ over a commutative field k, every element $f \in k\langle X\rangle$ can be written in the form

$$f = a_0 + \sum x_i f_i, \quad \text{where } a_0 \in k, f_i \in k\langle X\rangle,$$

and a_0 and f_i are uniquely determined by f.

Proof

The form (5.19) of the element f is unique, since there are no relations between the x's, while the scalars can be moved about freely. Therefore by collecting up all terms with a left-hand factor x_1 we obtain an expression of the form $x_1 f_1$, with $f_1 \in k\langle X\rangle$; similarly for $x_2, \dots, x_n$ and the only term that remains is the constant term; by construction all these terms are unique. ■

The expression f_i is called the **right cofactor** of x_i in $x_i f_i$, and the mapping $f \mapsto f_i$ is the **right transduction** for x_i.

We also note the following result, characteristic of free algebras:

Theorem 5.11

Let $k\langle X\rangle$ be the free algebra in a set X over a commutative field k. Given any k-algebra A, any mapping $\varphi : X \to A$ can be extended in just one way to a homomorphism from $k\langle X\rangle$ to A.

Proof

We have seen that every element of $k\langle X\rangle$ has the form

$$f = \sum_{i_1 \dots i_r} a_{i_1 \dots i_r} x_{i_1} \dots x_{i_r}.$$

where the x_i are in X. If a homomorphism $\bar{\varphi}$ extending φ is to exist, it must satisfy

$$f\bar{\varphi} = \sum_{i_1 \dots i_r} a_{i_1 \dots i_r} (x_{i_1}\varphi) \dots (x_{i_r}\varphi), \tag{5.20}$$

and it is therefore uniquely determined. Now it is easily verified that the mapping $\bar{\varphi}$ defined on $k\langle X\rangle$ by (5.20) is a homomorphism. ■

In Section 5.1 we saw that Ore domains include all Noetherian domains (Theorem 5.4); by contrast any non-Ore domain contains a subring which is free over a ring as coefficient domain, as we shall now show.

Theorem 5.12

Let R be an integral domain. Then R is a faithful K-algebra, where $K = \mathbb{Z}$ if R has characteristic zero and $K = \mathbb{F}_p$ if R has characteristic p. Further, R is either an Ore domain or it contains a free K-algebra of infinite rank.

Proof

We have seen in Section 1.2 that R has characteristic 0 or a prime, and the first assertion is an immediate consequence of this fact. Suppose now that R is not right Ore, say $a, b \in R^\times$ are such that $aR \cap bR = 0$. We claim that the K-algebra A generated by a and b is free on these generators. For if not, then we have a relation $f(a,b) = 0$, where f is a polynomial in a and b. Let us take f of least degree; collecting the terms with a left factor a and those with a left factor b, we can write it as $f = \lambda + af_1 + bf_2$, where $\lambda \in K$ and f_1, f_2 are not both zero, for otherwise λ would be zero and f would be the zero polynomial. Assume that $f_1 \neq 0$; then $f_1(a,b) \neq 0$ by the minimality of deg f, but $f(a,b) = 0$, hence $f(a,b)b = 0$ and so

$$af_1(a,b)b = b(\lambda - f_2(a,b)b).$$

But this contradicts the assumption that $aR \cap bR = 0$, and it shows A to be free on a and b; now the subalgebra generated by $a^n b$ ($n \in \mathbb{N}$) is free of countable rank. ∎

For polynomial rings in one variable the division algorithm is a useful tool, which has no known analogue valid for more than one variable. However, in the non-commutative case such an analogue exists and may be defined as follows.

Let R be a ring with a degree function; thus for each $a \in R^\times$ a non-negative integer $\mathrm{d}(a)$ is defined, while $\mathrm{d}(0) = -\infty$, such that

D.1 $\mathrm{d}(a) \in \mathbb{N}$, for $a \neq 0$, while $\mathrm{d}(0) = -\infty$;
D.2 $\mathrm{d}(a-b) \leq \max\{\mathrm{d}(a), \mathrm{d}(b)\}$;
D.3 $\mathrm{d}(ab) = \mathrm{d}(a) + \mathrm{d}(b)$.

We shall introduce a notion of dependence on R related to the degree, which will play an important role in what follows. A family (a_i) $(i \in I)$ of elements of R is said to be **right** d-**dependent** if some a_i vanishes or there exist elements $b_i \in R$, almost all zero, such that

$$\mathrm{d}\Big(\sum a_i b_i\Big) < \max\{\mathrm{d}(a_i b_i)\}. \tag{5.21}$$

Clearly the family (a_i) must contain more than one element (unless it reduces to 0). It may well be infinite, but in any case the sum in (5.21) is defined, because almost all the b_i are 0. In particular, any right linearly dependent family is also right d-dependent. For if a family (b_i) exists such that $\sum a_i b_i = 0$, where not all the b_i vanish, but no a_i is 0, choose a term, $a_i b_i$ of highest degree, r say; then the sum of all the terms $a_j b_j$ of degree r has degree less than r, so (5.21) holds. The family (a_i) is called **right** d-**independent** if it is not right d-dependent.

Secondly, we shall call an element $a \in R$ **right** d-**dependent** on the family (a_i) if either $a = 0$ or there exist $c_i \in R$, almost all 0, such that

$$\mathrm{d}\left(a - \sum a_i c_i\right) < \mathrm{d}(a) \quad \text{and} \quad \mathrm{d}(a_i c_i) \leq \mathrm{d}(a) \quad \text{for all } i. \tag{5.22}$$

If a is not right d-dependent on (a_i), it will be called **right** d-**independent** of (a_i). Of course there are also the notions of **left** d-**dependence**, defined in a corresponding way. We collect some elementary properties of this dependence:

Proposition 5.13

Let R be a ring with a degree function d. Then (i) any right d-independent family is right linearly independent, (ii) any element right d-dependent on a given family is also d-dependent on a finite subfamily. Moreover, this dependence notion is transitive: if a is d-dependent on $a_1, \ldots, a_n$ and each a_i is d-dependent on $u_1, \ldots, u_m$ then a is d-dependent on $u_1, \ldots, u_m$.

Proof

Since any right linearly dependent family is right d-dependent, (i) follows. To prove (ii), we note that any d-dependence relation involves only finitely many elements, so the first part of (ii) follows. Suppose now that a is d-dependent on the a's and each a_i on the u's:

$$\mathrm{d}\left(a - \sum a_i c_i\right) < \mathrm{d}(a), \quad \mathrm{d}(a_i c_i) \leq \mathrm{d}(a),$$

$$\mathrm{d}\left(a_i - \sum u_k h_{ki}\right) < \mathrm{d}(a_i), \quad \mathrm{d}(u_k h_{ki}) \leq \mathrm{d}(a_i).$$

Let us write $a_i = a_i^* + \sum u_k h_{ki}$ and note that $\mathrm{d}(a_i^*) < \mathrm{d}(a_i)$. We have

$$a - \sum a_i c_i = a - \sum (a_i^* + u_k h_{ki}) c_i = a - \sum u_k \left(\sum h_{ki} c_i\right) - \sum a_i^* c_i,$$

hence

$$\mathrm{d}\Big(a - \sum u_k\Big(\sum h_{ki}c_i\Big)\Big) < \mathrm{d}(a), \quad \mathrm{d}\Big(u_k\Big(\sum h_{ki}c_i\Big)\Big) \leq \mathrm{d}(a),$$

where we could omit the terms in a_i^* because they are of lower degree. This shows a to be right d-dependent on the u's, as claimed. ■

Using these notions, we shall now define a form of division algorithm for free algebras. The main difference between it and the usual commutative form is that it involves families of elements satisfying a certain condition (which is vacuous in the commutative case and the case of a one-element family). A ring R with a degree function is said to possess the **weak algorithm** if in any finite right d-dependent family of non-zero elements $a_1, \ldots, a_n$, where $\mathrm{d}(a_1) \leq \ldots \leq \mathrm{d}(a_n)$, some a_k is right d-dependent on $a_1, \ldots, a_{k-1}$. Thus there exist $c_1, \ldots, c_{k-1}$ such that

$$\mathrm{d}\left(a_k - \sum_{i=1}^{k-1} a_i c_i\right) < \mathrm{d}(a_k), \quad \mathrm{d}(a_i c_i) \leq \mathrm{d}(a_k).$$

A family containing a finite subfamily with this property is also called **strongly right d-dependent**. The weak algorithm can then be expressed by saying that every right d-dependent family is strongly right d-dependent.

Let R be a ring with a degree function d satisfying the weak algorithm. We claim that the elements of non-positive degree form a subfield of R. These elements form a subring of R, by D.2–D.3. Further, if $\mathrm{d}(a) = 0$, then $a.1 - 1.a = 0$, hence 1 is right d-dependent on a, i.e. there exists $b \in \mathbb{R}$ such that $\mathrm{d}(1 - ab) < 0$, hence $ab = 1$. It follows that $\mathrm{d}(b) = 0$ and so there exists $c \in \mathbb{R}$ such that $bc = 1$. Thus $c = abc = a$, and b is the inverse of a. In this way we obtain

Proposition 5.14

In any ring with a degree function satisfying the weak algorithm the elements of degree zero form together with 0 a subfield (possibly skew) of R. ■

To justify the introduction of the weak algorithm we shall now show that it holds in free algebras; in fact, with another fairly plausible condition it characterizes free algebras:

Theorem 5.15

Let R be a ring which is also a k-algebra, for some field k, with a degree function d such that the set of elements of degree zero is $k^{\times}$. Then R satisfies the weak algorithm relative to d if and only if $R = k\langle X\rangle$ for some set X.

Proof

Assume that $R = k\langle X\rangle$, with a degree function which assigns a positive degree to each $x \in X$; we have to show that R satisfies the weak algorithm. We remark that R as k-space has a basis consisting of all **monomials**, i.e. products of elements of X. For any monomial $m = x_1 \dots x_s$ of positive degree r we define the **left transduction** for m as the linear mapping $a \mapsto a^*$ of R into itself which maps any monomial bm to b and all other monomials to 0. Thus a^* is the "left cofactor" of m in the canonical expression for a, and by Proposition 5.10 it is uniquely determined by a. Clearly we have $\mathrm{d}(a^*) \leq \mathrm{d}(a) - r$, and for any $a, b \in R$, if b is a monomial of degree at least r, then either $b = b'm$ and so $b^* = b'$, $(ab)^* = ab'$, or b does not have m as right factor and $b^* = (ab)^* = 0$; in either case $(ab)^* = ab^*$. If $\mathrm{d}(b) < r$, then all we can say is that $\mathrm{d}((ab)^* - ab^*) < \mathrm{d}(a)$, thus for any monomial b,

$$\mathrm{d}((ab) * - ab^*) < \mathrm{d}(a). \tag{5.23}$$

By linearity this holds when b is any element of R.

To prove the weak algorithm, let $a_1, \dots, a_n$ be a right d-dependent family:

$$\mathrm{d}\left(\sum a_i b_i\right) < d = \max\{\mathrm{d}(a_i b_i)\}.$$

By omitting terms of degree less than d we may assume that $\mathrm{d}(a_i b_i) = d$ and we may assume the a_i numbered so that $\mathrm{d}(a_1) \leq \dots \leq \mathrm{d}(a_n)$, and so $\mathrm{d}(b_1) \geq \dots \geq \mathrm{d}(b_n)$. Our aim will be to show that some a_i is right d-dependent on the preceding terms.

Let $m = x_1 \dots x_s$ be a monomial of maximal degree $r = \mathrm{d}(b_n)$ occurring in b_n with a non-zero coefficient and denote the left transduction for m by *. In the expression $\sum a_i b_i^*$ the i-th term differs from $(a_i b_i)^*$ by a term of degree $< \mathrm{d}(a_i) \leq \mathrm{d}(a_n)$, because of (5.23). Hence the sum differs from $(\sum a_i b_i)^*$ by a term of degree less than $\mathrm{d}(a_n)$; but this sum has degree $\leq \mathrm{d}(\sum a_i b_i) - r < d - r = \mathrm{d}(a_n)$. Hence $\mathrm{d}(\sum a_i b_i^*) < \mathrm{d}(a_n)$ and since b_n^* is of degree 0 and so in $k^{\times}$, this is a right d-dependence relation of a_n on $a_1, \dots, a_{n-1}$.

Conversely, assume that R satisfies the weak algorithm relative to a degree d. To show that R is a free algebra we begin by constructing a free generating set X. Let us write R_d for the subspace of R consisting of all elements of degree at most d; thus $k = R_0 \subseteq R_1 \subseteq R_2 \subseteq \ldots, \bigcup R_n = R$. Further, let S_d be the subspace of R_d spanned by all products ab of degree at most d such that $\mathrm{d}(a), \mathrm{d}(b) < d$, and choose X_d as a basis for the complement of S_d in R_d. We claim that $X = \cup X_d$ is the required free generating set of R.

In the first place X is right d-independent, for if $x \in X$ of degree r is right d-dependent on the rest, say

$$\mathrm{d}\Big(x - \sum x_j b_j\Big) < r, \quad \mathrm{d}(x_j b_j) \leq r,$$

then $x - \sum x_j b_j \in R_{r-1}$, while any terms $x_j b_j$ with $\mathrm{d}(x_j) < r$ lie in S_r; it follows that $x - \sum x_j \beta_j \in S_r$, where $\beta_j \in k$ and $\beta_j = 0$ unless $\mathrm{d}(x_j) = r$. But this contradicts the fact that X_r is linearly independent mod S_r and it proves X to be right d-independent. It follows that X is linearly independent; moreover, an induction shows that the monomials in X of degree at most r span R_r. To complete the proof we show that the monomials are linearly independent. Thus assume a dependence relation $\sum x_I \alpha_I = 0$, where I ranges over the different index sets. In each term let us separate out the left-hand factor in X; then we obtain

$$\sum x a_x + \alpha = 0, \quad \text{where } x \in X, a_x \in R, \alpha \in k.$$

By the d-independence of X each $a_x = 0$, hence α must also vanish; now an induction on the degree shows that the given relation was trivial. Hence the different monomials in X are linearly independent and so form a k-basis for R, therefore R is the free k-algebra on X. ■

This result shows also that the weak algorithm for k-algebras as in Theorem 5.15 is left–right symmetric. This statement can be proved quite generally for any rings with a degree function, but that result will not be needed here (cf. Cohn, 1985, Section 2.3).

The free algebras defined above are not the most general types of ring, since firstly the ground field may be non-commutative and secondly it may not be centralized by the generators. These objections can be overcome by introducing the tensor ring on a bimodule. One can show that such rings again satisfy the weak algorithm; to develop their theory would take us too far afield, but we shall at least give the definition. Let K be any skew field with a central subfield k. By a K_k-**bimodule** we shall understand a K-bimodule such that the left and right actions by k agree. As an example we have the free K_k-bimodule, formed as follows: the tensor product $K \otimes_k K$ is a K_k-bimodule in a natural way, with the **canonical generator** $1 \otimes 1$. The direct sum of n copies of this module, in which the canonical generators are written $x_1, \ldots, x_n$, will be called the free K_k-bimodule on $x_1, \ldots, x_n$. Given any K_k-bimodule M, we define the r-th power as $M^r = M \otimes M \otimes \ldots \otimes M$ (with r factors) and form the K-ring, called the **tensorK_k-ring**

$$K_k\langle M\rangle = K \oplus M \oplus M^2 \oplus \ldots,$$

with multiplication induced by the natural isomorphism $M^r \otimes M^s \to M^{r+s}$. By assigning to each element f as degree the highest power r such that M^r contains a non-zero component, we obtain a degree-function and it can be shown that $K_k\langle M\rangle$ satisfies the weak algorithm relative to this degree function (cf. Cohn, 1985, Section 2.6).

Exercises 5.3

1. Show that if a is right d-dependent on a set B, then it is right d-dependent on the members of B of degree at most $\mathrm{d}(a)$. Show also that the right d-dependence is unaffected by adjoining or removing 0 from B.

2. Let R be a ring; a **filtration** on R may be defined as a series of subgroups of the additive group of R:
$$0 = R_{-\infty} \subseteq R_0 \subseteq R_1 \subseteq \ldots \subseteq R_n \subseteq \ldots,$$
such that (i) $\bigcup R_n = R$, (ii) $R_iR_j \subseteq R_{i+j}$, (iii) $1 \in R_0$. Define a function δ on R by
$$\delta(x) = \min\{n \mid x \in R_n\}.$$
Verify that $\delta(xy) \leq \delta(x) + \delta(y)$, and that equality holds here iff δ is a degree function.

3. State and prove an analogue of Gauss's lemma (Theorem 3.13) for $\mathbb{Z}\langle X\rangle$, the free algebra on a set X over $\mathbb{Z}$.

4. Show that in defining right d-dependence of an element a on a right d-independent family (a_i) the second condition in (5.22) can be omitted.

5. Factorize the element $xyzyx + xyz + zyx + xyx + x + z$ of $k\langle x, y, z\rangle$ in all possible ways. Verify that each factorization has the same number of terms, and compare the factors.

6. Use the weak algorithm for the free algebra $R = k\langle X\rangle$ to show that every invertible matrix over R is a product of elementary and diagonal matrices.

7. Give examples of non-free subalgebras of a free algebra. (Hint: try the 1-generator case first.)

8. Let $R = k\langle x, y\rangle$, where k is a field of characteristic zero. Given $a, b \in R$, if $ab + ba = 0$, show that $a = 0$ or $b = 0$.

9. Let $R = k\langle X\rangle$ be a free algebra and $C = [R, R]$ the ideal generated by all commutators $[a, b] = ab - ba (a, b \in R)$. Verify that R/C is a polynomial ring on X.

10. Let k, X, R, C be as in Exercise 9, where $|X| > 1$. Verify that $S = k + C$ is a subalgebra of R and show that S cannot be free. (Hint: expand $c[ac, bc]$, where $c = [a, b]$ and consider $[S, S]$.)

11. Let $R = k\langle X\rangle$ be a free algebra. If for each product u in X an element a_u of k is defined, the expression $f = \sum a_u u$ is called a **formal power series**. Show that the set $P = k\langle\langle X\rangle\rangle$ of all such expressions can be defined as a ring in a natural way (the **formal power series ring**), containing R as a subring. Show that P is a local ring.

5.4 Free Ideal Rings

We have seen in Chapter 3 that the polynomial ring in one variable over a field, $k[x]$, has a division algorithm and in Section 5.3 we met a generalization to several non-commuting variables, the weak algorithm. But we also saw that $k[x]$ is a principal ideal domain and we naturally ask whether this also has an analogue. Of course the free algebra $k\langle X\rangle$ is not a PID as soon as X has more than one element, but it is a free ideal ring in the sense of the following definition.

A ring R is said to be a **free right ideal ring**, briefly a **right fir**, if every right ideal of R is free, as right R-module, with a uniquely determined rank; of course the last condition just means that such a ring always has IBN. **Left firs** are defined similarly, and a **free ideal ring** or **fir** is a left and right fir. Later in this section we shall see that every free algebra is a fir, but we first have to establish some general properties of firs. In the first place, one often has to deal with a weaker condition, where only the finitely generated left or right ideals are free (still of unique rank). The resulting ring is called a **semifir**; such rings can also be defined by a condition on their relations. Consider a relation of the form

$$x_1y_1 + x_2y_2 + \ldots + x_ny_n = 0, \tag{5.24}$$

in a ring R. The relation (5.24) is called **trivial** if for each $i = 1, \ldots, n$, either $x_i = 0$ or $y_i = 0$. We can write (5.24) more briefly as $x.y = 0$, where x is a row vector $(x_1, \ldots, x_n)$ and y a column vector $(y_1, \ldots, y_n)^{\mathrm{T}}$. Now the relation $x.y = 0$ is said to be **trivializable**, if the relation $xP.P^{-1}y = 0$ is trivial, where P is a suitably chosen invertible matrix over R. For example, in any non-trivial ring we have a non-trivial relation

$$x.y = (1 \quad 1)\begin{pmatrix} -1 \\ 1 \end{pmatrix} = 0.$$

which can be trivialized by the matrix $P = \begin{pmatrix} 1 & -1 \\ 1 & 1 \end{pmatrix}$. More generally, if we have a matrix relation

$$XY = 0, \tag{5.25}$$

where X is $r \times n$ and Y is $n \times s$, then (5.25) is **trivial** if for each $i = 1, \ldots, n$, either the i-th column of X or the i-th row of Y is zero; if $XP.P^{-1}Y = 0$ is trivial for some invertible $n \times n$ matrix P, then (5.25) is said to be **trivialized** by P, and is called **trivializable**. The above example shows that every ring (apart from 0) has non-trivial relations, but as we shall now show, the rings in which every relation is trivializable are precisely the semifirs:

Theorem 5.16

For any non-zero ring R the following conditions are equivalent:

(a) R is a semifir;

(b) R has IBN and every finitely generated submodule of a free right R-module is free;

(c) every matrix relation in R can be trivialized;

(d) every relation in R can be trivialized.

Proof

(a) $\Rightarrow$ (b). Since R is a semifir, by (a), it has IBN. Now let F be a free right R-module and G a finitely generated submodule. The finite generating set of G involves only finitely many generators of F and by omitting the rest we may take F to be of finite rank n, say. Denote the projection of F on the first factor R by π_1 and put $F' = \ker \pi_1$. By restricting π_1 to G we obtain a finitely generated right ideal A of R and hence have the short exact sequence

$$O \to F' \cap G \to G \to A \to 0. \tag{5.26}$$

Since R is a semifir (by (a)), A is free, so the sequence splits and we have

$$G \cong (F' \cap G) \oplus A.$$

By induction on $n, F' \cap G$ as a finitely generated submodule of F', a free right R-module of rank $n-1$, is free, as well as A and it follows that G is free.

(b) $\Rightarrow$ (c). Let $XY = 0$ be a matrix relation over R, where X is $r \times n$ and Y is $n \times s$. The matrix X defines a linear mapping of column spaces by left multiplication, $f : {}^nR \to {}^rR$, so we have the short exact sequence

$$0 \to \ker f \to {}^nR \to \operatorname{im} f \to 0. \tag{5.27}$$

Here im f is a finitely generated submodule of ${}^rR$ and so is free; therefore (5.27) splits and by changing the basis in ${}^nR$ we obtain a basis adapted to the submodule ker f:

$${}^nR = \ker f \oplus F, \quad \text{where } F \cong \operatorname{im} f.$$

If P is the invertible matrix describing this change of basis, we put $X' = XP$, $Y' = P^{-1}Y$. Then $X'.Y' = X.Y = 0$ and so the columns of Y' lie in ker f. If t is the rank of ker f, then the rows of Y' after the first t are zero, while the first t columns of X' are zero, so the relation $XY = 0$ has been trivialized by P.

(c) $\Rightarrow$ (d) is clear and to prove (d) $\Rightarrow$ (a), take a finitely generated right ideal A of R and let n be the least integer such that A has an n-element generating set $u_1, \dots, u_n$, say. We have to show that A is free, so assume that there is a relation for the row vector $u = (u_1, \dots, u_n)$, say $u.a = 0$, where $a \neq 0$. By (d) this relation can be trivialized, say $u' = uP$, $a' = P^{-1}a$. Since $a \neq 0$, we have $a' \neq 0$, say $a'_n \neq 0$, but $u'.a' = 0$ is trivial, so $u'_n = 0$ and A is generated by $u'_1, \dots, u'_{n-1}$, which contradicts the choice of n, and this proves that A is free on $u_1, \dots, u_n$. If A has another basis $v_1, \dots, v_m$, where $m \neq n$, then $m > n$; consider the endomorphism f of A defined by mapping $v_1 \mapsto u_1, \dots,$ $v_n \mapsto u_n, v_{n+1} \mapsto 0, \dots, v_m \mapsto 0$. Since $v_m f = 0$, f is not injective, but it is surjective, so $v_1 f, \dots, v_n f$ generate A, but not freely, and it follows by what was shown earlier, that A can be generated by fewer than n elements, which is again a contradiction; hence $m = n$ and A has a uniquely determined rank. ■

This result shows that left and right semifirs are the same, by the symmetry of the condition (d), so we need only speak of semifirs. Returning to firs, we now come to an important source of firs:

Theorem 5.17

Let R be a ring with a degree function d, satisfying the weak algorithm. Then every right ideal of R has a right d-independent basis; hence it is free and the rank is unique. Similarly for left ideals, so that R is a fir.

Proof

The argument here is quite similar to that used to prove that every ideal in a Euclidean ring is principal (Theorem 3.14). Define $R_n = \{x \in R | \mathrm{d}(x) \leq n\}$; then $K = R_0$ is the subfield of elements of non-positive degree. Given any right ideal I of R, we construct a basis of I as follows. We may assume that $I \cap K = 0$, since otherwise I would contain a unit and so be the whole ring. Put $I_n = I \cap R_n$ and assume that for $n < r$ a basis B_n of each I_n has been constructed such that $C_r = B_1 \cup B_2 \cup \ldots \cup B_{r-1}$ is right d-independent. Denote by I'_r the subset of elements of I_r that are right d-dependent on C_r; clearly I'_r contains I_{r-1} and is closed under addition and right multiplication by K, i.e. it is a K-space, which corresponds to a subspace I'_r/I_{r-1} of I_r/I_{r-1}. We pick a K-basis for a complement of this subspace in I_r/I_{r-1} and take the inverse images of this basis in I_r. The set B_r of these elements together with C_r, denoted by C_{r+1}, is a right d-independent generating set of I_r, for any d-dependence would lead to a linear dependence over K of the basis we have chosen. This proves the induction step; thus by induction on r we obtain a d-independent generating set of I and this shows I to be a free right ideal. Since at each step the number of basis elements is uniquely determined (as the dimension of a certain K-space), the rank of I is unique and this shows R to have IBN. Thus R is a right fir; if we assume the fact (proved in Cohn, 1985, Corollary 2.3.2) that the weak algorithm is symmetric, the same argument, applied to left ideals, shows R to be a left fir, and hence a fir, as claimed. ■

As mentioned in Section 5.1, a general ring may well have several non-isomorphic fields of fractions. In the general theory it is shown that for a certain class of rings (which includes all semifirs) there is a unique **universal field of fractions**, from which all other fields of fractions can be obtained by specialization. A detailed account would take us too far afield (see Cohn, 1985, 1995), but we shall conclude by showing that a free algebra has several non-isomorphic fields of fractions:

Theorem 5.18

Let k be a commutative field and $R = k\langle x, y\rangle$ the free k-algebra on x, y. Then there are countably many skew field extensions of R that are pairwise non-isomorphic over R.

Proof

Consider the polynomial ring $P = k[t]$, with the endomorphism α_n mapping t to t^n. The twisted polynomial ring $P[x;\alpha_n]$ is right but not left Noetherian, and therefore by Theorem 5.12 contains a free k-algebra, in fact x and $y = xt$ form a free generating set. Let R be the algebra generated by x, y; it is a subalgebra of $P[x;\alpha_n]$ which, being right Noetherian, has a skew field of fractions, $K^{(n)}$ say. Thus R has been embedded in a skew field, and there is no isomorphism between $K^{(n)}$ and $K^{(m)}$ (for $m \neq n$) over R, because each $K^{(n)}$ contains $t = x^{-1}y$, but $y^{-1}x^{-1}yx = t^{n-1}$. ■

Exercises 5.4

1. Determine which of the following are semifirs or firs, where k is a commutative field: k, $k[x]$, $k[x, y]$, $\mathbb{Z}\langle x, y\rangle$, $k[x^a \,|\, a \in \mathbb{Q}^+]$.

2. Let K be a ring and R, S subrings that are semifirs. Is $R \cap S$ necessarily a semifir?

3. Show that a right fir is a right Ore domain iff it is a PID. Find the conditions for a semifir to be a right Ore domain.

4. Show that the formal power series ring $k\langle\langle X\rangle\rangle$ defined in Exercise 5.3.11 is a semifir.

5. Let R be a semifir, F a free right R-module and let A, B be finitely generated submodules of F. By considering the natural homomorphism $A \oplus B \to A + B$ show that $\mathrm{rk}(A + B) + \mathrm{rk}(A \cap B) = \mathrm{rk}\, A + \mathrm{rk}\, B$.

6. Let R be a semifir and M a finitely presented R-module. For any presentation $M = F/G$ define $\chi(M) = \mathrm{rk}\, F - \mathrm{rk}\, G$ and show that this is independent of the choice of presentation (assumed finite). (Hint: use Schanuel's lemma.)

7. Let R be a semifir and $A \in {}^mR^n, B \in {}^nR^p$. If $AB = 0$, show that rk A + rk B $\leq$ n, where rk denotes the inner rank (see Section 1.2).

8. Let R, A, B be as in Exercise 7 but instead of $AB = 0$, AB now has an $r \times s$ block of zeros. Show that there is an invertible $n \times n$ matrix P and an integer $t, 1 \leq t < n$ such that AP has an $r \times t$ block of zeros and $P^{-1}B$ has an $(n - t) \times s$ block of zeros (this result is known as the **partition lemma**).

9. Let R, A, B be as in Exercise 7 (but AB is not prescribed). Using the definition of inner rank, show that $\text{rk } A + \text{rk } B \leq \text{rk } AB + n$. (This is known as Sylvester's law of nullity, first proved for matrices over a field in 1884 by James Joseph Sylvester, 1814–1897.)

10. Let $R = k\langle x, y\rangle$ be the free k-algebra on x, y and let α be the endomorphism defined by $x \mapsto x, y \mapsto xy$. Show that α is injective. If S is the minimal extension of R on which α extends to an automorphism, constructed as for Exercise 5.2.3, show that S is a semifir but not a left fir (it can be shown that S is a right fir, cf. Cohn, 1985, Section 2.10).

Outline Solutions

["finitely generated" is sometimes abbreviated as f.g.]

1.1

1. By R.6, $(a+1)(b+1) = a(b+1) + b + 1 = ab + a + b + 1$; expanding the second bracket first, we have $(a+1)b + a + 1 = ab + b + a + 1$, and $a + b = b + a$ follows by cancellation.
2. The unit element is (0,1).
3. For a ring homomorphism we require $ax + ay = a(x+y), ax.ay = axy$, $a.1 = 1$; hence we require $a = 1$.
4. Products with $0, 1$ are clear. $\mathbb{Z}/(3) = \{0,1,2\}$, where $2^2 = 1$, $\mathbb{Z}/(4) = \{0,1,2,3\}$, where $2^2 = 0, 2.3 = 3.2 = 2, 3^2 = 1$. The first is a field, but not the second.
5. Routine verification (try 2 factors first).
6. If the cyclic group has order m, this is just $\mathbb{Z}/(m)$.
7. Last part: $AB \subseteq A$, $AB \subseteq B$, hence $AB \subseteq A \cap B$. Now $1 = a + b$, where $a \in A, b \in B$, so if $c \in A \cap B$, then $c = aca + bca + acb + bcb \in AB$.
8. If $X \subseteq A_i$, then $X \subseteq \bigcap A_i$, hence $(X) \subseteq \bigcap A_i$, where (X) is the ideal generated by X. Since $X \subseteq (X)$, (X) is one of the A_i and we have equality.
9. (n). resp. $(n)/(m)$, where $n \mid m$.
10. If $x \in X, y \in \mathscr{C}_R(X)$, then $xy = yx$, so $x \in \mathscr{C}_R(\mathscr{C}_R(X))$; if X is commutative, then $xy = yx$ for all $x, y \in X$.

1.2

1. If $x.y = x + y + xy$, then $(x.y).z = x + y + xy + z + xz + yz + xyz = x.(y.z), x.0 = 0.x = x$ and when $x \neq -1$, then $x + y + xy = 0$ has the solution $-(1+x)^{-1}x$.
2. Elements: $0, 1, a, a^2$, where $a^3 = 1$. Addition: $a + a = 0$ (characteristic 2), $1 + a = a^2$ (it cannot be $0, 1$ or a).
3. If $a \in R$ is regular, then $na = n1.a = 0$ holds iff $n1 = 0$, so a has the same order as 1, and this can be infinite or any positive integer.
4. $aba = a$ in an integral domain implies $ba = 1$. For a counter-example in general take the monoid acting on infinite sequences $(u_1, u_2, \ldots)$ by the rules $(u_1, u_2, \ldots)a = (0, u_1, u_2, \ldots)$, $(u_1, u_2, \ldots)b = (u_2, u_3, \ldots)$ and take the ring generated by a and b.
5. If $aba = 1$, then ab is a left inverse and ba a right inverse of a, hence $ab = ba$ is the inverse.
6. Put $M_c = \{x \in M \mid x.1 = cx\}$, where $c = 0, 1$. Then M_0, M_1 satisfy V.1–V.3 and $x = (x - x.1) + x.1$, so $M = M_0 + M_1$; clearly $M_0 \cap M_1 = 0$.
7. Routine.
8. i, iii, iv, vi, vii.
9. Dimension $n + 1$ (basis:$1, x, x^2, \ldots, x^n$).
10. $1, x, x^2, \ldots$ are linearly independent.
11. If $\boldsymbol{v} = A\boldsymbol{u}$, then $\boldsymbol{v}' = Q\boldsymbol{v} = QA\boldsymbol{u} = QAP^{-1}\boldsymbol{u}'$. By reduction to Smith normal form A takes the form $I \oplus 0 = \text{diag}(1, \ldots, 1, 0, \ldots, 0)$. Since the rank remains unchanged, it is equal to the number of 1's.
12. A has rank m iff its rows are linearly independent, i.e. $XA = 0 \Rightarrow X = 0$. When this is so, the m rows of A span F^m, i.e. $AA' = I$ for some A',and the converse is clear. Moreover, the rank is then m, so $n \geq m$.

1.3

1. If $D_i(c)$ is the matrix with (i, i)-entry c and the rest zero, then for any diagonal matrix $A = \text{diag}(a_1, \ldots, a_n)$, $AD_i(c) = D_i(a_i c)$, $D_i(c)A = D_i(ca_i)$.
2. Routine.
3. If $Au = Bu$ for all u, then $Ae_i = Be_i$, i.e. the i-th columns of A and B agree. If $u_1, \ldots, u_n$ is a basis, then every vector u has the form $u = \sum l_i u_i$, hence $Au_i = Bu_i$ for all i implies $Au = Bu$.
4. Routine.
5. The determinant of a triangular matrix is the product of the diagonal elements.
6. $C = (c_{ij})$, where $c_{ij} = 0$ unless $i < j$, but there are only $n - 1$ steps from 1 to n, so any product of n or more c's is zero.

7. Take an invertible matrix P with first column u such that $Cu = 0$ (e.g. u could be a non-zero column of C^k, with the largest k such that $C^k \neq 0$). Then $P^{-1}CP$ has the first column zero; now use induction on n.
8. If $A = (a_{ij})$, $X = (x_{ij})$, then tr $AX = \sum_{rs} a_{sr}x_{rs} =$ tr XA. Taking $x_{ij} = 1$ and all other entries of X zero, we find tr $AX = a_{ji}$.
9. Put $e_{12} = u$, $e_{21} = v$, $e_{11} = uv$, $e_{22} = vu$ and check (1.14).
10. Routine.
11. $b^{i-1}ab^{1-j}b^{k-1}ab^{1-l} = \delta_{kj}b^{i-1}ab^{1-l}$. Rest clear.
12. Routine (try the 2×2 case first).

1.4

1. First part is routine. A can be defined as $\mathbb{Z}/(m)$-module, provided that A has exponent dividing m (i.e. $mA = 0$).
2. The condition for linear dependence of (a_1, a_2), (b_1, b_2) is: $b_i = \lambda a_i$ for some $\lambda \in \mathbb{Q}^\times$. Example: (4,6),(10,15).
3. Given $x \in M_S$ and $a \in R$, define $x.a = x(af)$.
4. Routine.
5. Routine.
6. The bottom row reads: $M/\ker f \cong \operatorname{im} f$; now use Theorem 1.21.
7. If M is generated by $X = X' \cup X''$, where X' is the finite subset of X involved in the defining relations, then $M = F' \oplus F''$, where F' is generated by X' with the defining relations for M and F'' is free on X''.
8. To get a left inverse for β, take a free generating set of $M''$and map them to their inverse images under β; any such mapping can be extended to a homomorphism.
9. Routine.
10. The direct sum of two cyclic groups of coprime orders r and s is cyclic of order rs.

1.5

1. Routine.
2. If f has an inverse g, then $fg = 1$, $gf = 1$; if g' is another inverse, then $g' = g'fg = g$.
3. For any ring R, the mapping $n \mapsto n.1_R$ is a homomorphism; it is determined by the image of 1 which must be 1_R, so it is unique. Similarly, $a \mapsto 0$ is the only homomorphism to 0.
4. The least, resp., greatest element.
5. Any skeleton of Ens has members of arbitrary cardinal and so cannot be small.
6. For each set S of n elements choose a fixed bijection from S to n and use it to translate mappings between sets.

7. If S, T, U are functors from I to A, and $f : S \to T$, $g : T \to U$ are natural transformations, then $fg : S \to U$ is a natural transformation from S to U and the identity has the properties of a neutral element.
8. $f : A \to B$ can be factorized as $f = gh$, where $g : A \to Z$, $h : Z \to B$, with zero object Z; then g, h are uniquely determined by A, B resp., by the definition of Z as zero object. Of the four categories only Gp has a zero object.
9. The composition of morphisms that include Z among its source or target is uniquely defined, by definition of Z.
10. The initial object is easily seen to be $1_P : P \to P$.

2.1

1. An endomorphism is A-linear and k is (isomorphic to) a subfield of A.
2. Routine.
3. If $a \in \mathscr{Z}(A)$ and $a \neq 0$, then $aA = A$, so $ab = 1$ for some b, which must be the inverse of a. For any $x \in A$, $x = abx = xab = axb$, hence $bx = xb$.
4. Routine. Substitute from (2.2) in (2.12).
5. Routine. If A has a unit element e, then in $A^1, e1 = 1e = e$.
6. By Theorem 2.1, A^1 is embedded in k_{n+1}, hence so is A.
7. Routine.
8. $(xy)ab = xab.y + x.yab + xa.yb + xb.ya$; when we form $[a, b]$, the last two terms cancel.
9. $(\exp d)^{-1} = \exp^{-d}$.
10. Associativity is clear. The rest follows from $(x \wedge y) \wedge z = (x \uparrow z)(S^2 + S) \uparrow y - (y \uparrow z)(S^2 + S) \uparrow x$.
11. Routine.
12. Routine.

2.2

1. $C_i / C_{i+1} \cong k^n$, as right k_n-module.
2. $A + (B \cap A) = (A + B) \cap A$ by Theorem 2.6; A contains the left- and is contained in the right-hand side.
3. If f is injective but not surjective on M, then im f is a proper submodule isomorphic to M, giving rise to a strictly descending sequence im f^n. For a surjective but not injective g, ker g^n is a strictly ascending chain.
4. n must be a prime power.
5. If $f_i : M_i \to M_{i+1}$, then $\ell(M_i) - \ell(M_{i+1}) = \ell(\ker f_i) - \ell(\ker f_{i+1})$; now form the alternating sum.
6. Given $N \subset M$, take composition series of N and M/N.
7. The index $(\mathbb{Z} : C_i)$ grows without bound, so any n fails to lie in some C_i.
8. Use Theorem 2.6.

9. If $(A+B)\cap C = A\cap C + B\cap C$, then $(A+C)\cap(B+C) = A\cap(B+C)+C\cap(B+C) = A\cap B + A\cap C + C\cap B + C \subseteq A\cap B + C$. The opposite inclusion is clear.
10. If $f: P\to M$, then $N\cap Pf = Nf^{-1}f$ for any submodule N of M, hence $(A+B)f^{-1}f = (A+B)\cap Pf = A\cap Pf + B\cap Pf = Af^{-1}f + Bf^{-1}f$, hence $(A+B)f^{-1} = (A+B)f^{-1}ff^{-1} \subseteq Af^{-1} + Bf^{-1}$. For the converse apply the inclusion mapping of C in M. The other assertion follows by duality.

2.3

1. Each row is a simple right ideal and all are isomorphic. S^k is free iff $r|k$.
2. First part routine. Second part: Just one, containing a simple component of each A_i.
3. By semisimplicity, $R = A\oplus B$ for some right ideal B. If $1 = e+f$, $e\in A$, $f\in B$, then for any $x\in A$, $x = ex+fx$, hence by uniqueness, $ex = x$, $fx = 0$. By symmetry there exists e' such that $xe' = x$ for all $x\in A$, hence $e' = e$ is the unit element of A.
4. R_n is a direct sum of n isomorphic R_n-modules, each with the same right ideal structure as R, so each one is Artinian, hence so is R_n.
5. If $A = (a_{ij})$ is in the centre of R, then $Ae_{rs} = e_{rs}A$, hence $\Sigma_i a_{ir}e_{is} = \Sigma_j e_{rj}a_{sj}$. The only term on both sides is $a_{rr}e_{rs} = a_{ss}e_{rs}$, hence $A = aI$, where $a\in \mathscr{Z}(R)$.
6. If R is simple, it has the form K_n for a skew field K, and its centre (by Exercise 5) is the centre of K, a field. Otherwise it is a direct product of more than one matrix ring and its centre is a direct product of more than one field.
7. If (2.27) is the decomposition of R into minimal ideals, any simple right R-module is isomorphic to a minimal right ideal in one of the ideals, so there are just t of them, up to isomorphism.
8. Write the elements of M^n as rows and operate with R_n on the right. M^n is cyclic iff M has an n-element generating set.
9. S is a sum of simple modules, hence semisimple. Any left multiplication by an element of R is an endomorphism, hence the image is again a sum of minimal right ideals, and so is contained in S.
10. First part clear. Second part: eR is a two-sided ideal iff e is central.

2.4

1. If A is a nilideal in an Artinian ring R with radical N and $x\in A$, then xa is nilpotent for all $a\in R$, so xR is a nilideal mod N, hence $xR\subseteq N$.

2. Any matrix A in N is upper 0-triangular and zero outside an $n \times n$ square, for some n, hence $A^n = 0$. But $C_n = e_{nn+1} \in N$ and $C_1C_2...C_{n-1} = e_{1n} \neq 0$.
3. If $R/N = A_1 \oplus \ldots \oplus A_r$, where the A_i are ideals that are simple as rings, and B_i is the sum of all A_j except A_i, let C_i be the ideal of R corresponding to B_i; then $N = \cap C_i$.
4. $\mathbb{R}$ as $\mathbb{Q}$-space is infinite-dimensional; let $u_1, u_2, \ldots \in \mathbb{R}$ be linearly independent over $\mathbb{Q}$, denote by U_n the subspace spanned by $u_1, \ldots, u_n$ and let V_n be a complement of U_n such that $V_n \subseteq V_{n-1}$. Then (U_n) is an ascending and (V_n) a descending chain. Further, $\mathbb{R}$ as $\mathbb{R}$-module is of length 1; the rest is routine.
5. If A is a minimal right ideal and $A^2 \neq 0$, then $cA \neq 0$ for some $c \in A$. By minimality, $cA = A$, so $c = cu$ for some $u \in A$. $c(u^2 - u) = 0$, so if N is the right annihilator of c, then $c(N \cap A) = 0$, but $cA = A$, hence $N \cap A = 0$ and so $u^2 = u$. Thus u is an idempotent in A and clearly $uR = A$.
6. Each minimal right ideal has the form e_iR, where e_i is an idempotent, and R is the direct sum of all the e_iR. Hence $e_j = \sum e_ie_j$ and by directness, $e_ie_j = 0 (j \neq i)$. Hence $u = \sum e_i$ is idempotent and $1 - u$ annihilates everything, so u is a left one. Similarly there is a right one, and we know they must be equal.
7. If $f = 1 - e$, then $x = exe + exf + fxe + fxf$.
8. Let R have radical N; for each simple component S_i of R/N take an idempotent e_i generating S_i and take a Peirce decomposition for $\sum e_i$.
9. Observe that N^i/N^{i+1} as R-module is annihilated by N.
10. Show by induction that $N^{r-i} \subseteq S^i(R)$.

2.5

1. m/n is a non-unit iff $p|m$; there are no other local subrings.
2. Each composition series yields a direct decomposition.
3. If M is decomposable, $\mathrm{End}_R(M)$ contains an idempotent $\neq 0, 1$.
4. The idempotents are 1, 16, 40, 81, 96, 105.
5. Decompose M, N, N' and apply the Krull–Schmidt theorem. Use the same method for Exercise 6.
7. Any decomposition of M leads to an idempotent (by decomposing 1). $\mathbb{Z}$ is indecomposable.
8. Let B be an indecomposable ideal of R. Then B is a direct summand and so has a central idempotent e; by decomposing $(1 - e)R$ find a complete decomposition. Now $B \cong A_i$ for some i, by the Krull–Schmidt theorem.
9. If the cyclic submodules of M are totally ordered and N is generated by $u_1, \ldots, u_r$, let Ru_i be the largest of $Ru_1, \ldots, Ru_r$; then $N = Ru_i$.

Conversely, if every 2-generator submodule is cyclic and Ra, Rb are given, then $Ra + Rb = Rc$, hence $a = xc$, $b = yc$, $c = ua + vb$, so $c = uxc + vb$. If x is a non-unit, then $1 - ux$ is a unit, so $Ra \subseteq Rc \subseteq Rb$. Otherwise $Rb \subseteq Rc \subseteq Ra$.

10. Regard K^n as a $K[x]$-module with $ux = Au(u \in K^n)$ and apply Fitting's lemma.

2.6

1. Imitate the proof of Theorem 2.35.
2. For equality ρ must be orthogonal.
3. Routine.
4. (u, v) is a unitary invariant positive-definite form, so any submodule has an orthogonal complement that is again a submodule.
5. Routine.
6. Apply Theorem 2.33 to reduce V.
7. If $f = \sum_{x \in G} x$, then $fy = yf(y \in G)$ and $f^2 = |G| f = 0$.
8. The eigenvalues of S are ± 1; since det $S = 1$, both are the same sign, so $S = \pm I$, but I was excluded. If $S^r = I$, the eigenvalues are roots of $x^r = 1$, say α, α^{-1} (because det $S = 1$); now a reduction to triangular form gives the result. A further transformation reduces b to 0.
9. Take $\rho(a)$ upper and $\rho(b)$ lower triangular; their product has characteristic equation $T^2 - cT + I = 0$, where c, determined by the non-diagonal terms of $\rho(a)$, $\rho(b)$, is at our disposal and so can be chosen to ensure that the eigenvalues are r-th roots of 1.
10. det : $G \to \mathbb{C}$ is a homomorphism with kernel N, so G/N is isomorphic to a finite subgroup of $\mathbb{C}$, which must be cyclic.

2.7

1. There are $(G : G') = 4$ linear characters; besides the trivial character, those taking on i, j, k, the values $-1, -1, 1; -1, 1, -1; 1, -1, -1$. The remaining representation has degree $2 = (8 - 4)^{1/2}$, with characters 0 on $\pm$i, $\pm$j, $\pm$k, and ± 2 on ± 1.
2. The derived group is $G' = \{1, (12)(34), (13)(24), (14)(23)\}$, so there are $3 = (G : G')$ linear characters. The conjugacy classes are 1; $G' \setminus \{1\}$; $\{(123), (134), (142), (243)\}$ and the inverse of this class, so the non-linear character has degree $(12 - 3)^{1/2} = 3$. Let G act on u_1, u_2, u_3, u_4 by permutations and take the submodule generated by $u_1 - u_2$. This gives the character $3, -1, 0, 0$.
3. Routine.
4. Express the characters in terms of irreducible characters and note that the latter form an orthonormal system (cf. (2.65)).

5. $(A : B)$.
6. Apply Theorem 2.45.
7. The character of ρ is a multiple of the character of the regular representation.
8. The operation $\rho(b) : a \mapsto eab$ represents projection on I followed by the right regular representation and $a.\rho(b) = \sum e(uv^{-1}w^{-1})a(w)b(v)u$. Using G as a basis, we have $x\rho(b) = \sum \rho_{xy}(b)y$; now put $a = x$, $u = y$.
9. Any central element is a linear combination of $c_1, \ldots, c_r$, so the latter form a basis of the centre. The rest follows by multiplying out.
10. Transform $\rho(c_1)$ to diagonal form and take traces. For the last part form the characteristic equation of the matrix $(g_{lmn})_{ln}$.

3.1

1. If A is generated by $c_1, \ldots, c_r$, the mapping $x_i \mapsto c_i$ extends to a homomorphism j from $F = k[x_1, \ldots, x_r]$ to A which is surjective; any ideal of A corresponds to an ideal of F containing ker j and so is finitely generated.
2. Imitate the proof for the degree.
3. Same proof as Theorem 3.3 but with order replacing degree.
4. Any non-zero element has the form $f = c(1 - gx)$, where $c \neq 0$ and g is a power series; its inverse is $c^{-1} \sum (gx)^n$.
5. $f = a_0 + a_1 x + \ldots + a_n x^n$ is a unit if a_0 is a unit and $a_1, \ldots, a_n$ lie in the maximal ideal of R.
6. An element f of $R[x]$ is invertible iff mod $J[x]$ it is an invertible matrix.
7. Let J be the unique maximal ideal of L; then $L_n/J_n = (L/J)_n$.
8. If the maximal ideal of L is J, then the element $a_0 + a_1 x + \ldots + a_n x^n$ is a unit iff a_0 is a unit.
9. If V is spanned by $x_1, \ldots, x_n$, then R is spanned by all power products $x_1^{i_1} \ldots x_n^{i_n}$; now imitate the proof of Theorem 3.3.
10. Arrange a total ordering of the power products by divisibility and choose $g_1 \in A$ so as to have the least leading term. If $g_1, \ldots, g_n$ have been chosen, take $g_{n+1} \in A \setminus (g_1, \ldots, g_n)$ with least possible leading term.

3.2

1. Choose q with $\theta(a - bq)$ minimal; if $\theta(a - bq) \geq q(b)$, then for some $q', \theta(a - bq - bq') < \theta(a - bq)$, contradicting the minimality of the latter.
2. If $f = gq + r = gq' + r'$, where r, r' have degree less than g, then $g(q - q') = r' - r$, which is a contradiction if $r'. \neq r$.
3. If $k > \log a / \log 2$, then $2^k > a$. Now $r_1 < \min(b, a - b) \leq a/2$; by induction, $r_k < a/2^k < 1$.
4. If $a, b \in R^\times$, then $\theta(a) = \theta(b)$ and for some $q, \theta(a - bq) < \theta(a)$; hence $a = bq$ and so $R^\times$ is a group.

5. $x^n - c^n = (x-c)(x^{n-1} + x^{n-2}c + \ldots + c^{n-1})$, hence $f(x) - f(c) = (x-c)q$. So $(x-c) \mid f$ iff $f(c) = 0$.
6. Straightforward induction.
7. The n-th Fibonacci number (the series c_n, where $c_0 = c_1 = 1$, $c_{n+1} = c_n + c_{n-1}$).
8. Any invertible matrix over R must have a unit in each row and so we can reduce all other entries in that row to zero by elementary transformations.
9. $\begin{pmatrix} 0 & 1 \\ -1 & 0 \end{pmatrix} = B_{12}(1)B_{21}(-1)B_{12}(1)$; without the sign change this is impossible because the determinant would be -1.
10. Routine.

3.3

1. Let p be an atom and $p \mid ab$; either $p \mid a$ or $pr + as = 1$ for some r, s. In the latter case $b = prb + abs$, and this is divisible by b. The conclusion follows by Theorem 3.11.
2. Routine.
3. If (3.21) are complete factorizations of a, b and $a_i = b_i$ for $i = 1, \ldots, k$ while the other factors are non-associated, then $d = a_1 \ldots a_k (= b_1 \ldots b_k)$, $m = ab_{k+1} \ldots b_s$, hence $ab = md$.
4. Let m be an LCM of a and b and put $d = ab/m$. Then $b \mid m$, hence $ab/m \mid a$; similarly $ab/m \mid b$; suppose $c \mid a$, $c \mid b$; then $m \mid ab/c$, hence $c \mid ab/m$, i.e. $c \mid d$.
5. Use induction: $\mathrm{HCF}(a_1, \ldots, a_n) = \mathrm{HCF}(a_1, \mathrm{HCF}(a_2, \ldots, a_n))$; similarly for LCM. For infinite subsets the HCF is still defined, but the LCM may not be (e.g. take the set of all prime numbers in $\mathbb{Z}$).
6. Every element of R involves finitely many of the x's and so lies in $k[x_1, \ldots, x_n]$ for some n.
7. Routine.
8. If $c = aa' = bb'$, where a is an atom not dividing b or b', then $a \mid bb'$, but a is prime to b and to b'.
9. If a, b are right associated, then $a = bc$, $b = ad$, hence $a = bc = adc$, so $dc = 1$, by symmetry $cd = 1$, so c, d are units.
10. The proof of the last part of Theorem 3.11 does not assume commutativity.

3.4

1. The parallelogram rule shows that a/d is associated to m/b.
2. If there are only finitely many atoms, say $a_1, \ldots, a_n$, then $a_1 \ldots a_n + 1 \notin k$, so it is a non-unit and hence divisible by an atom, which must clearly be different from the a_i.

3. Let $u = a/b$ and $Ra + Rb = Rc$; then $a = a'c, b = b'c$, where a', b' are coprime; hence $u = a'/b'$ and a', b' are unique up to a common unit factor.
4. Under the natural homomorphism $R \to R_T$ ideals of R_T correspond to ideals of R, so R_T is again Noetherian. Given an ideal A of R_T, we can bring the finite generating set to a common denominator, so $A = (a_1/b, \ldots, a_n/b)$. If $\mathrm{HCF}(a_1, \ldots, a_n) = (c)$, then $A = (c/b)$.
5. Let A be an ideal in R and take an ascending chain of finitely generated, hence principal ideals in A; this must break off, but it can do so only when it reaches A.
6. $(2x, x^2)$ is not principal and the generators have no LCM.
7. 1, 2, 20.
8. Two invariant factors $\neq 1$ are equal.
9. A is associated to $\mathrm{diag}(f_1, \ldots, f_r, 0, \ldots, 0)$; choose $c \in F$ such that $f_1(c) \ldots f_r(c) \neq 0$.
10. Apply Theorem 3.18 to make a PAQ-reduction of A to diagonal form.

3.5

1. Given $x \in M$, suppose that $xa = 0$ and let $a = q_1 \ldots q_r$ be the factorization of a into powers of different primes, say q_i is a power of p_i. Put $s_i = a/q_i$; the s_i have no common factor, so $\sum s_i c_i = 1$ for some c_i, hence $x = \sum x s_i c_i$ and $x s_i c_i q_i = 0$, so $x s_i c_i \in M_{p_i}$. A routine argument proves directness.
2. Use Theorem 3.23 and apply Exercise 1 to the torsion submodule.
3. Any submodule is torsion-free and if finitely generated is principal, hence $\cong R$.
4. (x, y).
5. As in Theorem 3.18 we can reduce the first row to $(a_1, 0, \ldots, 0)$ by right multiplication by an invertible matrix, and an induction now completes the reduction. The diagonal elements are unique up to right associates, provided A is non-singular.
6. $xR \cong R/cR$ for some $c \in R$; hence $xR \cong c^{-1}R/R \subseteq K/R$. This homomorphism $xR \to K/R$ can be extended to M, for details see the proof of Theorem 4.22.
7. If $a = bc$, then $R/Rb \cong Rc/Ra \subseteq R/Ra$, so we have the exact sequence $0 \to R/Rb \to R/Ra \to R/Rc \to 0$.
8. Use Exercise 7 and induction on the number of cyclic summands.

3.6

1. diag(1, 3).
2. $\mathbb{Q}(\mathrm{i}) : \pm 1, \pm \mathrm{i}$. $\mathbb{Q}(\sqrt{-d}) : \pm 1$.

3. This is the ring of integers in the field generated over $\mathbb{Q}$ by a cube root of 1, i.e. a root $\neq 1$ of $x^3 = 1$, or $x^2 + x + 1 = 0$, or also $4x^2 + 4x + 4 = 0$, giving $\mathbb{Q}(\sqrt{-3})$.
4. If $I = (a)$, put $J = (a^{-1})$; if $IJ = (c)$, then $c^{-1}J$ is an inverse.
5. $J = (1, (1 - \sqrt{-5})/2)$.
6. (x, y); any element a of the inverse J must be such that xa and ya are integral, hence a is integral and so $(x, y)J \subseteq (x, y)$, so $J \subseteq R$, a contradiction.

4.1

1. 47, −13.
2. If R' is another such ring, then there are homomorphisms $R \to R'$ and $R' \to R$ which are mutually inverse (a universal construction).
3. $x \in A_i$ if the i-th component is zero; now (i), (ii) follow.
4. $\mathbb{Z}$ satisfies the conditions on S in Proposition 4.3.
5. (i) provides a homomorphism $\mathbb{Z} \to \prod \mathbb{Z}/(p)$, which is injective because no non-zero integer is divisible by all primes; (ii) yields a mapping $\mathbb{Z} \to \prod \mathbb{Z}/(2^n)$, which is injective because no non-zero integer is divisible by all powers of 2.
6. The eigenvalues are 0 or 1, and they cannot be both 0 or both 1, so one is 0 and one is 1; hence tr $C = 1$, det $C = 0$.
7. Define $f = (f_i)$, acting on the components; f followed by projection on R_i is just f_i, hence it is surjective. ker $f_i = A_i$, hence ker $f = \cap$ ker $f_i = \cap A_i$. If $A_i + A_j = R$, let C_i be the intersection of all A_j except A_i; then $A_i + C_i = R$, so there exists $e_1 \in R$ mapping to $(1, 0, \ldots, 0)$ and similarly for e_i. Now $f : \sum a_i e_i \mapsto (a_1 f_1, \ldots, a_n f_n)$, which is surjective.
8. Put $f = e + ex(1-e)$; then $f^2 = f$, so $f = ef = fe = e$, hence $ex(1-e) = 0$.
9. $[ex(1-e)]^2 = 0$, hence $ex(1-e) = 0$.

4.2

1. The set of all chains in a partially ordered set is clearly inductive, so Zorn's lemma yields a maximal chain. Conversely, the union of a maximal chain gives a maximal element.
2. If the i-th set is $\{a_{ij}\}$, observe that there are $n-1$ terms a_{ij} with $i + j = n$, so order the a_{ij} according to increasing $i + j$.
3. The terms of the subset can be counted by taking them in the natural order.
4. If the i-th term is $0.a_1a_2\ldots$, form $0.b_1b_2\ldots$, where $b_i = a_i + 1$ if $a_i \neq 8$ or 9, and $b_i = 1$ otherwise.
5. Choosing a member of each set amounts to choosing an element of the product.
6. In each set pick the least number (for both cases).

7. Let C be the family of linearly independent subsets and verify (i) that C is inductive, and (ii) that a maximal linearly independent subset spans the whole space.
8. (b) $\Rightarrow$ (c) and (c) $\Rightarrow$ (d).
9. The commutative subrings containing A form an inductive family, so Zorn's lemma can be used.
10. Let F be a maximal proper subgroup; there exist $a, b \in \mathbb{Z}$ such that $a \in F$, $1/b \notin F$; by maximality $\mathbb{Q} = \{m/b\} + F$, so $1/ab^2 = m/b + c/d$, where $c/d \in F$; hence $1 = mab + ab^2c/d \in F$, so $1/b = ma + abc/d \in F$, a contradiction.

4.3

1. Verify that the right-hand side satisfies the universal property for tensor products.
2. $A \otimes A \cong A$, $(A \oplus B) \otimes (A \oplus B) \cong (A \otimes A) \oplus (B \otimes B)$.
3. Both sides represent the set of all bilinear mappings from $U \times V$ to W.
4. The mapping is clearly additive and for any $a \in R$, $b \in R^{\times}$, $xab^{-1} = (x \otimes 1)ab^{-1} = xa \otimes b^{-1}$. If x is a torsion element, say $xb = 0$, then $x \otimes 1 = xb \otimes b^{-1} = 0$ and conversely (cf. Proposition 5.5).
5. The functor $M \mapsto M_K$ preserves exact sequences, hence it is exact. For the given exact sequence $M/M' \cong M''$; this still holds after passage to M_K, so $\mathrm{rk}(M) = \mathrm{rk}(M') + \mathrm{rk}(M'')$.
6. This follows by linearity, factoring out $\det(A)$ from $\det(A \otimes B)$.
7. Linearity is clear and $(x \otimes 1)(y \otimes 1) = (xy \otimes 1)$. If $A \otimes B$ is associative, write down the associative law with 1 for the elements of B, to prove the associativity of A.
8. $u \otimes v$ commutes with $x \otimes y$ iff u commutes with x and v with y.
9. If $2U = 0$ and V is spanned by $v_1, \dots, v_n$, put $y_i = 2v_i$; then the given relation is non-zero in $U \otimes V'$, but zero in $U \otimes V$.
10. If (4.20) holds in $U \otimes V$, it also holds in $U \otimes V'$,where V' is a free submodule of V containing the y_i.

4.4

1. Let $f : x \mapsto \langle x, f \rangle \in \mathrm{Hom}_R(M, N)$; then S- and T-actions, defined by $\langle x, sf \rangle = \langle xs, f \rangle, \langle x, ft \rangle = \langle x, f \rangle t$, commute, but there is no way of defining a natural R-action.
2. If $\mathscr{J}(R)M = M \neq 0$, let M' be a maximal submodule of M; then M/M' is simple, so $\mathscr{J}(R)M/M' = 0$, hence $\mathscr{J}(R)M \subseteq M' \neq M$, a contradiction.
3. Apply Exercise 2 to M/N.
4. $\mathscr{J}(R) = (x)$ and $x^n \neq 0$ for all n.
5. A is in either side iff $I - XAY$ is invertible, for all X, Y.

6. Suppose that $\mathcal{J}(R) \not\subset A$ and take $c \in \mathcal{J}(R) \setminus A$, then the ideal generated by c consists of quasi-regular elements, so R/A cannot be semiprimitive.
7. Let $c \in R$ be such that $c + A$ is quasi-regular; then for some c', $c + c' + cc' \in A$, hence it is itself quasi-regular and it follows that c is quasi-regular. Hence there is a surjective homomorphism from $\mathcal{J}(R)$ to $\mathcal{J}(R/A)$, and its kernel is clearly A.
8. If $f \in \mathcal{J}(R[t])$, then for some $g \in R[t]$, $ft + g = fgt$ and a comparison of orders in t gives a contradiction.

4.5

1. If $m = m_1 \ldots m_n$, where the m_i are powers of primes, then any indecomposable projective has the form $\mathbb{Z}/(m_i)$ and any *f.g.* projective is a direct sum of such modules. A generator is of the form $\mathbb{Z}/(m')$, where m' is a multiple of all the m_i for different primes.
2. Let P be a *f.g.* projective and write $P \cong F/G$, where F is free of finite rank. Then $F \cong P \oplus G$; hence G, as homomorphic image of F, is *f.g.*
3. Given $c \in I$, we have $c = \sum a_i b_i c = \sum a_i (b_i c)$. Now observe that $c \mapsto (b_i c)$ is a linear mapping from I to R and apply the dual basis lemma.
4. Clearly a principal right ideal is free (of rank 0 or 1). Conversely, if a right ideal I is free on $a_1, a_2, \ldots$, then $a_1 a_2 - a_2 a_1 = 0$, a contradiction, which shows I to be principal.
5. Apply the dual basis lemma to the equation $\sin\theta(\sin\theta/2(1 - \cos\theta)) + \frac{1}{2}(1 - \cos\theta) = 1$; the ideal is not principal since $\sin\theta$ and $1 - \cos\theta$ have no common (rational) factor.
6. There is a surjection $R^n \to P$, which splits because P is projective. If the projections from R^n to P, P' are described by matrices E, E', then $E + E' = I$, $EE' = 0$, hence E is idempotent.
7. If P is presented by R^n, it is free iff there is a change of basis such that P is the isomorphic image of R^r; here r is the rank of P.
8. Over a local ring every idempotent matrix is similar to $I_r \oplus 0$.
9. Let $R/\mathcal{J}R) \cong K_n$; every idempotent matrix can be transformed so that mod $\mathcal{J}(R)$ it is of the form $I_r \oplus 0$, hence the n-th power is of the required form.
10. Take P to be the submodule of elements of $M_1 \oplus M_2$ whose components have the same image in N.
11. Form the pullback and apply Exercise 10.

4.6

1. Take S to be $N_1 \oplus N_2$ modulo the submodule identifying the two images of any element of M. The rest is routine.
2. Form the pushout and apply Exercise 1.

3. For each $x \neq 0$ in M take a copy I_x of I and a mapping $M \to I_x$ which is non-zero on x; the corresponding mapping $M \to \prod I_x$ is an embedding.
4. Every cyclic module has the form $\mathbb{Z}/(m)$ and this can be embedded in $\mathbb{Q}/\mathbb{Z}$ by mapping 1 to $1/m$.
5. Any injective submodule of I is a direct summand; by Zorn's lemma there is a least injective submodule containing M and a routine verification shows that it is a maximal essential extension of M.
6. The resolution is $\ldots \to \mathbb{Z} \to \mathbb{Z} \to \mathbb{Z}/(4) \to (2) \to 0$.
7. Clearly R is injective and we have the resolution $0 \to (2) \to R \to R \to \ldots$, so $\text{cdim}(2) = \infty$.

4.7

1. If IBN fails for R, then $AB = \mathrm{I}$ for a pair of rectangular matrices; now apply the homomorphism.
2. Routine.
3. Suppose R is of type (m, p); thus $R^m \cong R^{m+p}$. If M is generated by n elements, it is a homomorphic image of R^n, but $R^n \cong R^{n-rp}$, where r is chosen so that $m \leq n - rp < m + p$. Hence $h = m + p - 1$.
4. If for each n there is a *f.g.* module which cannot be generated by n elements and if R^n is presented by R^m, then $m \geq n$. Thus $R^m \cong R^n \oplus K$ implies $m \geq n$, which translates into the stated matrix condition.
5. $P = A'B, Q = B'A$.
6. Any element of R can be written as a sum of products of a's and b's, and the degree can be reduced by multiplying by suitable b's on the left and a's on the right.
7. By hypothesis, $R^m \cong R^n$ for some $m \neq n$; now take endomorphism rings.

5.1

1. Ore condition for a monoid M: given $a, b \in M, aM \cap bM \neq \emptyset$. The proof is a simplified form of that of Theorem 5.1.
2. Let $ab^{-1} \in K$, $c \in R$ and $bc' = cb'$. Then $ab^{-1}c = ac'b'^{-1}$. Given a linearly dependent subset of K, multiply by a common denominator and use the resulting relation to eliminate one of the elements.
3. Use the exactness of localization and the relation $\dim M_K = \dim M'_K + \dim M''_K$. For a longer exact sequence (M_i), $\sum(-1)^i \dim M_i = 0$.
4. Clearly $c\alpha$ must be non-zero for $c \neq 0$; when this is so, the unique extension is given by $(ab^{-1})\alpha = (a\alpha)(b\alpha)^{-1}$.
5. Verify that Ore subrings form an inductive set.
6. If $a, b \neq 0$ and $aR + bR = dR$, then $a = da_0, b = db_0$, and $a_0u - b_0v = 1$ for some $u, v \in R$. Hence $a(ua_0 - 1) = bva_0$, and this is easily shown to be non-zero.

7. If ab^{-1} is in the centre, it commutes with b, so $a = bab^{-1}$, and $ab^{-1}x = xab^{-1}$, hence $axb = bxa$; the converse is clear.
8. If a, b are right large, then for any $c \neq 0$, there exist $a', c' \in R$ such that $ac' = ca' \neq 0$, and $b'', c'' \in R$ such that $bc'' = c'b''$, hence $abc'' = ac'b'' = ca'b''$, so $abR \cap cR \neq 0$.
9. R might have $AB = cI$, for rectangular A, B and c is a non-unit in S.
10. $[E : F] < \infty$.

5.2

1. To find $(b^{-1})\delta$, we have $0 = 1.\delta = (bb^{-1})\delta = b\delta.(b\alpha)^{-1} + b.(b^{-1})\delta$. The rest is a routine verification.
2. Routine.
3. $K^\alpha \cong K$ and $K^\alpha \subseteq K$, so we can embed K in a field isomorphic to K, K' say. Now $K \subseteq K' \subseteq K'' \subseteq \ldots$, and the union is a field containing K, on which α is an automorphism. When $R = K(x; \alpha)$, then $K' = xKx^{-1}$.
4. $(fg)^\alpha = f^\alpha g^\alpha$, hence $(fg)T = fT.g^\alpha$, so $gT = Tg^\alpha$.
5. Use the formula $uv = vu^v$, where $u \in N$, $v \in L$ and $u^v = v^{-1}uv$.
6. $(x + a)R + (y + b)R \ni c = [x + a, y + b]$, hence the sum is the whole ring. Denoting these ideals by I, J we have the short exact sequence $0 \to I \cap J \to I \oplus J \to R \to 0$. This sequence splits and each of I, J is free of rank 1, hence $I \cap J \oplus R$ is free; if $I \cap J$ were free, it would have to have rank 1 (by IBN) and so be principal, which is clearly not the case.
7. $\alpha^2 = 1$, $\alpha\delta = \delta$.
8. The subfield of the centre of K fixed by α.
9. Routine.

5.3

1. If $\mathrm{d}(a - \sum b_i c_i) < \mathrm{d}(a)$ and some b_i have degree greater than $\mathrm{d}(a)$, then the sum of terms in $\sum b_i c_i$ of degree greater than $\mathrm{d}(a)$ vanishes and so may be omitted.
2. Routine.
3. A product of primitive polynomials is primitive, with the same proof as Theorem 3.13.
4. Under the given conditions $\mathrm{d}(a_i c_i) \leq \max\{\mathrm{d}(a_i c_i)\} = \mathrm{d}((\sum a_i c_i - a) + a) \leq \mathrm{d}(a)$.
5. $xyzyx + xyz + zyx + xyx + x + z = (xyz + x + z)(yx + 1) = (xy + 1)(zyx + x + z)$.
6. Adapt the proof of Theorem 3.9.
7. $k\langle x\rangle$: subalgebra generated by x^2 and x^3. $k\langle x, y\rangle$: subalgebra generated by k and all commutators.

8. Suppose $a, b \neq 0$; add a relation $yx = nxy$, with $n \in \mathbb{N}$ to be chosen later. If α denotes the endomorphism $y \mapsto ny$, and $f = \sum x^i a_i$, $g = \sum x^j b_j$, then the leading term in $ab + ba$ has coefficient $a_m \alpha^n b_n + a_m (b_n) \alpha^m$ and this does not vanish for suitable n.
9. R/C is just the ring R made abelian.
10. $c[ac, bc] = ca[c, bc] + cb[ac, c] + c^4$.
11. Routine verification.

5.4

1. $k, k[x]$ are firs, $k[x^a \mid a \in \mathbb{Q}^+]$ is a semifir.
2. Put $F = k\langle x, y, u, v \mid uv = vu = 1\rangle$ (a fir, by Cohn, 1985, Section 7.10, though this fact will not actually be used). Now define $x_i = xu^i$, $y_i = v^i y$ $(i \in \mathbb{Z})$ and consider the subalgebra A generated by the x_i, y_i and u and the subalgebra B generated by the x_i, y_i and v. A is a semifir, as a union of the firs A_n generated by x_{-n}, y_n and u; similarly for B. However, $A \cap B$ is not a semifir, since the relation $x_0 y_0 - x_1 y_1 = 0$ cannot be trivialized in $A \cap B$.
3. If a right fir is right Ore, every right ideal is free of rank at most 1. A semifir is right Ore iff it is right Bezout.
4. Verify the weak algorithm relative to the (negative) order function.
5. Note that the short exact sequence $0 \to A \cap B \to A \oplus B \to A + B \to 0$, is split exact.
6. Routine verification.
7. Apply the semifir condition (c) of Theorem 5.16.
8. Apply the same condition to the $r \times n$ row block of A and the $n \times s$ column block of B whose product is zero.
9. Write $AB = PQ$, where Q has the least number of rows, namely $\text{rk}(AB)$. Then $(A \quad P)\begin{pmatrix} B \\ -Q \end{pmatrix} = 0$, so by Exercise 7, we have $\text{rk}(AB) + n \geq \text{rk}(A \quad P) + \text{rk}(B \quad -Q)^{\mathrm{T}} \geq \text{rk}\, A + \text{rk}\, B$.
10. Clearly x and xy are free generators; as a direct limit of semifirs, S is a semifir, but not a fir, because the left ideal generated by x, $x^{-1}y$, $x^{-2}y, \ldots$ is not free.

Notations and Symbols used Throughout the Text

■ indicates the end of a proof (or if the latter is easy, its absence)

(a) $\Rightarrow$ (b)	(a) implies (b)
(a) $\Leftrightarrow$ (b)	(a) is equivalent to (b)
$x \in S$	x is a member of the set S
$\{x \in S \mid P(x)\}$	the set of all members x of S satisfying $P(x)$
$A \subseteq B$ or $B \supseteq A$	A is contained in B
$A \subset B$ or $B \supset A$	A is a proper subset of B
$A \cap B$	intersection of A and B
$A \cup B$	union of A and B
$B \setminus A$	set of members of B not in A
$\varnothing$	the empty set
$\mathscr{P}(S)$	the collection of all subsets of S
$a \leq b$ or $b \geq a$	a is less than or equal to b
$a < b$ or $b > a$	a is strictly less than b
$\max(a_1, \ldots, a_r)$	the greatest a_i
$\min(a_1, \ldots, a_r)$	the least a_i
$A \to B$	mapping from A to B
$a \mapsto b$	mapping which takes a to b
$\mathbb{N}$	set of all natural numbers: $0, 1, 2, 3, \ldots$
$\mathbb{Z}$	set of all integers: $0, \pm 1, \pm 2, \pm 3, \ldots$
$\mathbb{Q}$	set of all rational numbers
$\mathbb{R}$	set of all real numbers
$\mathbb{F}_p$	field of p elements

$\mathbb{C}$	set of all complex numbers
$\mathbb{H}$	set of all quaternions
$a\,|\,b$	a divides b
$\mathbb{Z}/(m)$	set of all integers mod m
$a \equiv b \pmod{m}$	$m\,|\,a-b$
$\sum_i a_i$	sum over all a_i
$\prod_i c_i$	product over all c_i
A^I	set of all functions from I to A
$\mathrm{End}(A)$	set of all endomorphisms of A
$\mathrm{Hom}(A, B)$	(homo)morphisms from A to B
$R[x, y, \ldots]$	set of all polynomials in $x, y, \ldots$ over R
$\deg f$	degree of the polynomial f
$\mathfrak{M}_n(R)$	set of all $n \times n$ matrices over R , $\mathfrak{M}_n(R)$
δ_{ij}	Kronecker delta
$\mathscr{L}_F(U, V)$	set of all linear transformations from U to V
$\mathscr{L}(V)$	set of all linear transformations of V
$\boldsymbol{u}$	row or column of vectors
V^*	dual of V
$\langle f, g\rangle$, (f, g)	bilinear form in f, g
A^{T}	transpose of the matrix A
${}^mR^n$	set of all $m \times n$ matrices over R
e_{ij}	matrix unit
$\mathrm{tr}\, A$	trace of the matrix A
$\det A$	determinant of the matrix A
$\mathrm{diag}(a_1, \ldots, a_n)$	matrix with $a_1, \ldots, a_n$ on the main diagonal and all other entries zero
R°	ring opposite to R
G/N	quotient group of G by N
G'	commutator subgroup of G
$|G|$	order of the group G (also used for absolute value)
$\mathrm{im}\, f$	image of the mapping f
$\ker f$	kernel of the mapping f
$\mathrm{coker}\, f$	cokernel of the mapping f
$A \cong B$	A is isomorphic to B
$\mathscr{Z}(R)$	centre of the ring R
$\mathscr{C}_R(X)$	centralizer of X in R
$R^\times$	set of all non-zero elements of R
$R_1 \times R_2$, $\prod_i R_i$	direct product of the R_i
$A_1 \oplus A_2$, $\amalg_i A_i$	direct sum of the A_i
Ens	category of all sets and mappings
Gp	category of all groups and homomorphisms

Rg	category of all rings and homomorphisms
${}_R\mathscr{M}, \mathscr{M}_R$	category of all left, resp., right R-modules and homomorphisms
ρ_a	right multiplication by a
λ_c	left multiplication by c
$(x, y) = x^{-1}y^{-1}xy,$	also HCF of x and y
$[x, y] = xy - yx,$	also LCM of x and y
$\Delta(A)$	set of all derivations of A
$\ell(M)$	composition-length of the module M
$\mathrm{soc}(R)$	socle of the ring R
$\mathrm{rad}(R)$, $\mathscr{J}(R)$	(Jacobson) radical of the ring R
$\mathrm{GL}_n(R)$	group of all invertible $n \times n$ matrices over R
$\mathrm{SL}_n(R)$	subgroup of $\mathrm{GL}_n(R)$ of all matrices of determinant 1
$\mathrm{GE}_n(R)$	subgroup of $\mathrm{GL}_n(R)$ generated by all diagonal and elementary matrices
$\mathbb{T} = \mathbb{R}/\mathbb{Z}$	group of all real numbers mod 1
$A^\wedge$	character group of A
$R[[x]]$	formal power series ring
$R((x))$	ring of formal Laurent series
$(a_1, \ldots, a_n)$	ideal generated by $a_1, \ldots, a_n$, also row vector
$\mathrm{t}M$	torsion submodule of M
$\mathrm{rk}\ M$	rank of the module M
$\mathrm{N}(\alpha)$	norm of the algebraic number α
$R \otimes S$	tensor product of R and S
$R[x; \alpha, \delta]$	skew polynomial ring
$K(x; \alpha, \delta)$	skew function field
$k\langle X \rangle$	free algebra on X over k
$K_k\langle X \rangle$	tensor k-algebra on X over K

Bibliography

Suggestions for Further Reading

Cohn, PM (1982, 1989, 1991) Algebra, 2nd edn, 3 vols. John Wiley & Sons, Chichester (referred to in the text as A.1, A.2, A.3)
Fulton, W, Harris J (1991) Representation theory, a first course. Graduate Texts in Mathematics 129, Springer-Verlag, Berlin
Jacobson, N (1985, 1989) Basic algebra, 2nd edn, 2 vols. Freeman, San Francisco
Lam, TY (1992) A first course in noncommutative rings. Graduate Texts in Mathematics 131, Springer-Verlag, Berlin
Rotman, JJ (1979) An introduction to homological algebra. Pure & Appl. Math. 85, Academic Press, New York
van der Waerden, BL (1971, 1967) Algebra I, 8th edn; II, 5th edn. Springer-Verlag, Berlin

Other Works Quoted in the Text

Bass, H (1960) Finitistic dimension and a homological generalization of semi-primary rings. Trans. Amer. Math. Soc. 95: 466–488
Chase, SU (1960) Direct products of modules. Trans. Amer. Math. Soc. 97: 457–473

Cohn, PM (1966) On the structure of the GL_2 of a ring. Publ. Math. IHES 30, Paris

Cohn, PM (1985) Free rings and their relations, 2nd edn. London Math. Society Monographs 19, Academic Press, London

Cohn, PM (1995) Skew fields, theory of general division rings. Encyclopedia of Mathematics and its Applications, vol. 57, Cambridge University Press

Fraenkel, AA (1914) Über die Teiler der Null und die Zerlegung von Ringen, J. reine angew. Math. 145: 139–176

Fuchs, L (1958) Abelian groups. Hungarian Academy of Sciences, Budapest

Jacobson, N. (1996) Finite-dimensional division algebras over fields. Springer-Verlag, Berlin

Kaplansky, I (1972) Set theory and metric spaces, Allyn & Bacon, Boston

Nagata, M (1962) Local rings, Interscience Publishers, New York

Stewart, IN, Tall, DO (1987) Algebraic number theory, 2nd edn. Chapman & Hall, London

Wallace, DAR (1998) Groups, rings and fields. Springer-Verlag, London

Index